WHO Library Cataloguing-in-Publication Data

Water safety in distribution systems.

1.Water Quality. 2.Drinking Water. 3.Water Pollution. 4.Water Supply - standards. I.World Health Organization.

ISBN 978 92 4 154889 2 (NLM classification: WA 675)

© World Health Organization 2014

The illustration of the cover page is extracted from Rescue Mission: Planet Earth, © Peace Child international 1994; used by permission.

All rights reserved. Publications of the World Health Organization are available on the WHO website (www.who.int) or can be purchased from WHO Press, World Health Organization, 20 Avenue Appia, 1211 Geneva 27, Switzerland (tel.: +41 22 791 3264; fax: +41 22 791 4857; e-mail: bookorders@who.int).

Requests for permission to reproduce or translate WHO publications –whether for sale or for non-commercial distribution– should be addressed to WHO Press through the WHO website (www.who.int/about/licensing/copyright_form/en/index.html).

The designations employed and the presentation of the material in this publication do not imply the expression of any opinion whatsoever on the part of the World Health Organization concerning the legal status of any country, territory, city or area or of its authorities, or concerning the delimitation of its frontiers or boundaries. Dotted and dashed lines on maps represent approximate border lines for which there may not yet be full agreement.

The mention of specific companies or of certain manufacturers' products does not imply that they are endorsed or recommended by the World Health Organization in preference to others of a similar nature that are not mentioned. Errors and omissions excepted, the names of proprietary products are distinguished by initial capital letters.

All reasonable precautions have been taken by the World Health Organization to verify the information contained in this publication. However, the published material is being distributed without warranty of any kind, either expressed or implied. The responsibility for the interpretation and use of the material lies with the reader. In no event shall the World Health Organization be liable for damages arising from its use.

Design and layout: paprika-annecy.com

Printed by the WHO Document Production Services, Geneva, Switzerland

Contents

ACKNOWLEDGEMENTS 4

INTRODUCTION 6

WATER SAFETY PLANS 10

1. ASSEMBLE THE WSP TEAM 11

2. DESCRIBE THE WATER SUPPLY SYSTEM 13
   2.1 Types of water transmission or distribution system 14
   2.2 Components of water transmission or distribution systems 15
   2.3 Other important information and relevant performance indicators 16
   2.4 Informal settlements 17

3. IDENTIFY HAZARDS AND HAZARDOUS EVENTS AND ASSESS THE RISKS 19
   3.1 Hazards 20
      3.1.1 Microbial hazards 20
      3.1.2 Chemical hazards 22
         3.1.2.1 Disinfection by-products 22
         3.1.2.2 Chemicals from pipe materials and fittings 22
         3.1.2.3 Water treatment chemicals 23
      3.1.3 Physical hazards 24
   3.2 Hazardous events 24
      3.2.1 Physical integrity 26
         3.2.1.1 Physical faults, illegal connections, programmed shutdowns and recharge 26
         3.2.1.2 Cross-connections and backflow 27
         3.2.1.3 Contamination of storage reservoirs 28
         3.2.1.4 Plumbing issues and protecting the water supply at the customer interface 29
         3.2.1.5 Construction work, renovations and repairs 32
      3.2.2 Hydraulic integrity 33
         3.2.2.1 Changes in water flow due to intermittent water supply 33
         3.2.2.2 Changes in water flow due to mixing of water sources 34
      3.2.3 Water quality integrity 35
         3.2.3.1 Release of hazards from materials and fittings, including corrosion and scaling 35
         3.2.3.2 Treatment process changes with an impact on distribution 39
         3.2.3.3 Microbial growth and biofilms 41
         3.2.3.4 Water ageing 42
   3.3 Risk assessment 44
      3.3.1 Semi-quantitative risk assessment 44
      3.3.2 Quantitative microbial risk assessment 44

4. DETERMINE AND VALIDATE CONTROL MEASURES, REASSESS AND PRIORITIZE THE RISKS 47
   4.1 Determine current control measures 48
   4.2 Validation of control measures 52
   4.3 Reassess and prioritize the risks 54

5. DEVELOP, IMPLEMENT AND MAINTAIN AN IMPROVEMENT/UPGRADE PLAN 57

6. DEFINE MONITORING OF THE CONTROL MEASURES ... 59

- 6.1 Selection of appropriate operational monitoring parameters ... 60
  - 6.1.1 Types of parameters ... 64
  - 6.1.2 On-line operational monitoring ... 65
    - 6.1.2.1 Disinfectant residual ... 65
    - 6.1.2.2 Flow ... 65
    - 6.1.2.3 pH ... 65
    - 6.1.2.4 Pressure ... 65
    - 6.1.2.5 Temperature ... 65
    - 6.1.2.6 Turbidity ... 65
    - 6.1.2.7 Chemical parameters ... 66
    - 6.1.2.8 Biological monitoring devices ... 66
- 6.2 Reviewing operational monitoring data ... 66

7. VERIFY THE EFFECTIVENESS OF THE WSP ... 67

- 7.1 Verification monitoring ... 68
  - 7.1.1 Microbial parameters ... 69
  - 7.1.2 Chemical parameters ... 69
    - 7.1.2.1 Health-related chemicals ... 69
    - 7.1.2.2 Metals that influence acceptability ... 70
  - 7.1.3 Example distribution system verification monitoring programme ... 70
  - 7.1.4 Choosing sampling locations ... 70
    - 7.1.4.1 Designating sampling zones ... 70
    - 7.1.4.2 Selection of sampling sites ... 71
  - 7.1.5 Sampling frequency ... 71
- 7.2 Customer satisfaction ... 72
- 7.3 Internal and external auditing ... 72
  - 7.3.1 Internal audits ... 73
  - 7.3.2 External audits ... 73

8. PREPARE MANAGEMENT PROCEDURES ... 75

- 8.1 Standard operating procedures ... 76
  - 8.1.1 Positive pressure and adequate flows ... 77
  - 8.1.2 Intermittent flows ... 77
  - 8.1.3 Maintaining disinfectant residuals ... 77
  - 8.1.4 Mixing water sources ... 78
  - 8.1.5 Inspection and maintenance of storage tanks/service reservoirs, valves and other fittings ... 78
  - 8.1.6 Water leakage management ... 80
  - 8.1.7 Preventing corrosion ... 80
  - 8.1.8 Selection of pipe materials and chemicals ... 80
  - 8.1.9 Connection of new customers (including cross-connection control and backflow prevention) ... 81
  - 8.1.10 Repair of water main breaks ... 81
  - 8.1.11 Construction and commissioning of new mains ... 83
  - 8.1.12 Dewatering and recharging distribution mains ... 83
  - 8.1.13 Permeation ... 84
  - 8.1.14 Collection and testing of water samples (what, where, when, how and who) ... 84
  - 8.1.15 Calibrating equipment ... 84
  - 8.1.16 Customer complaints ... 84
- 8.2 Incident criteria and protocols ... 84

9. DEVELOP SUPPORTING PROGRAMMES ... 87

10. PLAN AND CARRY OUT PERIODIC REVIEW OF THE WSP ... 91

11. REVISE THE WSP FOLLOWING AN INCIDENT ... 93

12. ENABLING ENVIRONMENT 95
12.1 Regulatory and policy frameworks 96
12.1.1 Water quality legislation 96
12.1.2 Economic legislation 97
12.2 Independent surveillance, audits and inspections 98
12.3 Disease and outbreak surveillance 98
12.3.1 Disease surveillance 98
12.3.2 Disease and outbreak investigation 99
12.3.3 Communication, recording and reporting 100
12.3.4 Lessons learnt 100
12.4 Standards, codes of practice and certification 100
12.4.1 Distribution system design codes 100
12.4.2 National standards and certification systems 101
12.5 Capacity building 102
12.5.1 Training 102
12.5.2 Training providers 102
12.5.3 Maintenance contractors 102
12.5.4 Independent auditors 103
12.5.5 Risk assessors 103

REFERENCES 104

CASE-STUDY ANNEXES 115

Case-study 1:
Application of a predictive model for water distribution system risk assessment in India 115

Case-study 2:
Distribution network management utilizing the block system to reduce non-revenue water in Phnom Penh, Cambodia 120

Case-study 3:
Water safety plans and distribution systems – the case of Spanish Town, Jamaica 123

Case-study 4:
Drinking-water contamination incident in Johannesburg, South Africa 127

Case-study 5:
Incident management in the water distribution system in Johannesburg, South Africa 131

Case-study 6:
Black water complaints from western parts of Singapore in the 1980s 134

Case-study 7:
Implementation of massive replacement programmes for unlined galvanized iron connections and unlined cast iron mains in Singapore in response to poor water quality after commissioning of the Kranji/Pandan scheme (1983–1993) 137

Case-study 8:
Contamination of water supply incident in Bukit Timah Plaza/Sherwood Tower Condominium in Singapore in 2000 139

Case-study 9:
Introduction of the "TOKYO High Quality Program" (Tokyo's version of the water safety plan) 142

Case-study 10:
Incidents from "Lessons learnt from plumbing incidents – responses and preventions", Japan Water Plumbing Engineering Promotion Foundation (2011) 146

Acknowledgements

The World Health Organization (WHO) wishes to express its appreciation to all whose efforts made this production possible. In particular, WHO gratefully acknowledges the contributions of the following international experts, who contributed to and reviewed the publication:

Lead editor and advisor

David CUNLIFFE, South Australia Department of Health, Australia

Advisory team members

Jayant BHAGWAN, Water Research Commission, South Africa
Lesley D'ANGLADA, United States Environmental Protection Agency, United States of America
Asoka JAYARATNE, Yarra Valley Water, Australia
Nonhlanhla KALEBAILA, Water Research Commission, South Africa
Pawan LABHASETWAR, National Environmental Engineering Research Institute, India
Edward OHANIAN, United States Environmental Protection Agency, United States of America
Dai SIMAZAKI, National Institute of Public Health, Japan
Dave VIOLA, International Association of Plumbing and Mechanical Officials and World Plumbing Council, United States of America
Kee Wei WONG, PUB, the national water agency, Singapore

Authors

Nicholas ASHBOLT, School of Public Health, University of Alberta, Canada
David CUNLIFFE, South Australia Department of Health, Australia
Lesley D'ANGLADA, United States Environmental Protection Agency, United States of America
Peter GREINER, NSF International, United States of America
Rajesh GUPTA, Visvesvaraya National Institute of Technology, India
John HEARN, ALS Water Resources Group, Australia
Asoka JAYARATNE, Yarra Valley Water, Australia
Kah Cheong LAI, PUB, the national water agency, Singapore
Nicholas O'CONNOR, Ecos Environmental Consulting, Australia
Dave PURKISS, NSF International, United States of America
Irene TOH, PUB, the national water agency, Singapore
Dave VIOLA, International Association of Plumbing and Mechanical Officials and World Plumbing Council, United States of America
Kee Wei WONG, PUB, the national water agency, Singapore

Case-study contributors

Jayant BHAGWAN, Water Research Commission, South Africa
BUREAU OF WATERWORKS, Tokyo Metropolitan Government, Japan
Richard J. GELTING, Centers for Disease Control and Prevention, United States of America
JAPAN WATER PLUMBING ENGINEERING PROMOTION FOUNDATION, Japan
Pawan LABHASETWAR, National Environmental Engineering Research Institute, India
Mthokozisi NCUBE, Johannesburg Water, South Africa
Aabha SARGAONKAR, National Environmental Engineering Research Institute, India
Dai SIMAZAKI, National Institute of Public Health, Japan
Irene TOH, PUB, the national water agency, Singapore
WATER AND SEWER BUREAU, City of Kitakyushu, Japan
Kee Wei WONG, PUB, the national water agency, Singapore

Reviewers

Nicholas ASHBOLT, School of Public Health, University of Alberta, Canada
Michèle GIDDINGS, Health Canada, Canada
Kevin HELLIER, Melbourne Water, Australia
Darryl JACKSON, Consultant, Nepal
Claus JØRGENSEN, DHI, Denmark
Hamanth KASAN, Rand Water, South Africa
Mark W. LECHEVALLIER, American Water, United States of America
France LEMIEUX, Health Canada, Canada
Nicholas O'CONNOR, Ecos Environmental Consulting, Australia
Angella RINEHOLD, Consultant to the World Health Organization, United States of America
David SUTHERLAND, World Health Organization, India
Jakobus VAN ZYL, University of Cape Town, South Africa
Gordon YASVINSKI, Health Canada, Canada

The generous financial and technical support of the following is gratefully acknowledged: the United States Environmental Protection Agency; the Ministry of Health, Labour and Welfare of Japan; PUB, the national water agency, a statutory board under the Ministry of the Environment and Water Resources of Singapore; the Australian Department of Foreign Affairs and Trade; and the UK Department for International Development.

The development and production of this document were coordinated and managed by staff of the Water, Sanitation, Hygiene and Health (WSH) unit of WHO, including Mr Bruce Gordon (Coordinator), Ms Jennifer De France, Ms Sophie Boisson and Mr Kah Cheong Lai.

The professional editing services of Ms Marla Sheffer of Ottawa, Canada, and the secretarial support provided by Ms Penny Ward are also gratefully acknowledged.

Introduction

The integrity of well managed distribution systems is one of the most important barriers that protect drinking-water from contamination. However, management of distribution systems often receives too little attention. Distribution systems can incorrectly be viewed as passive systems with the only requirement being to transport drinking-water from the outlets of treatment plants to consumers.

There is extensive evidence that inadequate management of drinking-water distribution systems has led to outbreaks of illness in both developed and developing countries. The causes of these outbreaks and the range of chemical and microbial hazards involved are diverse. Between 1981 and 2010 in the United States of America (USA), 57 outbreaks were associated with distribution system faults, leading to 9000 cases of illness (CDC, 1981, 1982, 1983, 1984; St Louis, 1988; Levine, Stephenson & Craun, 1990; Herwaldt et al., 1991; Moore et al., 1993; Kramer et al., 1996; Levy et al., 1998; Barwick et al., 2000; Lee et al., 2002; Blackburn et al., 2004; Liang et al., 2006; Yoder et al., 2008; Brunkard et al., 2011; Hilborn et al., 2013). The most common faults were cross-connections and back-siphonage; other faults included burst or leaking water mains, contamination during storage, poor practices during water main repair and installation of new water mains, pressure fluctuations and leaching from pipework; a significant proportion of faults are unknown (Fig. 1(a)). Elsewhere, outbreaks of illness have been associated with low water pressure and intermittent supply (Hunter et al., 2005).

The most common causes of illness were enteric pathogens, including bacteria (*Salmonella*, *Campylobacter*, *Shigella*, *Escherichia coli* O157), protozoa (*Cryptosporidium*, *Giardia*) and viruses (Norovirus) (Fig. 1(b)). Chemicals, including copper, chlorine and lead, were associated with eight outbreaks (14%) (CDC, 1981, 1982, 1983, 1984; St Louis, 1988; Levine, Stephenson & Craun, 1990; Herwaldt et al., 1991; Moore et al., 1993; Kramer et al., 1996; Levy et al., 1998; Barwick et al., 2000; Lee et al., 2002; Blackburn et al., 2004; Liang et al., 2006; Yoder et al., 2008; Brunkard et al., 2011; Hilborn et al., 2013).

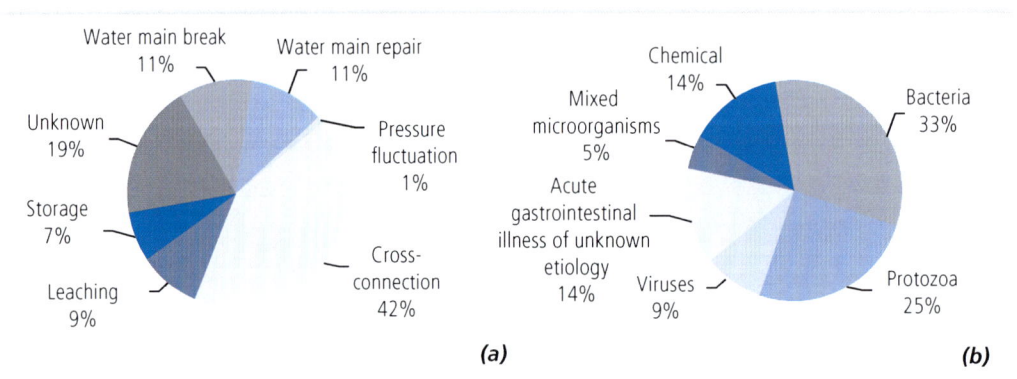

Fig. 1. Waterborne outbreaks associated with distribution systems in the USA, 1981–2010, by (a) system fault and (b) causative agent

Outbreaks are the most noticeable outcome of distribution system failures, but represent only a small fraction of contamination events. Cross-connections, leaks and water main breaks and transient low water pressures are common events; although most do not cause reported outbreaks, it is likely that some could cause sporadic cases of illness that go undetected by standard surveillance systems (Frost, Craun & Calderon, 1996). This is supported by a review of targeted epidemiological investigations that detected increased risks of gastrointestinal disease associated with distribution system deficiencies (Ercumen, Gruber & Colford, 2014).

A survey conducted in North America reported an average of seven breaks per 100 km of water main per year (Folkman, 2012); in Australia in 2011–2012, the rate was 13 breaks per 100 km of water main (National Water Commission, 2013). Well managed utilities have standard procedures for dealing with leaks and water main breaks; in the great majority of cases, breaks are repaired in the absence of reported outbreaks. In the Netherlands, 50 adverse water quality events, not associated with an outbreak, were reported between 1993 and 2004, based on repeated detection of coliforms or thermotolerant coliforms (Van Lieverloo et al., 2006). Over half of the events were associated with water main replacements or repairs. During this period, only one outbreak was detected, and this was caused by cross-connection between a drinking-water supply and partially treated river water (Van Lieverloo et al., 2006). There is no evidence that illness was caused by any of the other events, but it could not be completely ruled out.

In some circumstances, illnesses can arise in the absence of physical faults. Pathogenic microorganisms such as *Legionella* and the amoeba *Naegleria fowleri* can grow in biofilms, particularly in distribution systems that are susceptible to warming. *Naegleria fowleri* has caused deaths in Australia, Pakistan and the USA following nasal exposure to drinking-water containing the organism (Dorsch, Cameron & Robinson, 1983; Shakoor et al., 2011; CDC, 2014).

Inadequate management of distribution systems can be influenced by a range of factors, including a lack of understanding of the challenges that can threaten drinking-water safety during delivery to consumers. There can also be ownership and access issues. Cross-connections and backflow from buildings and facilities connected to distribution systems represent a common contamination threat – and have led to contamination of distribution systems (Jones & Roworth, 1996; USEPA, 2002a) – but water utilities typically have limited rights or access to ensure that the internal plumbing and storages within connected buildings provide adequate protection for their systems.

Intermittent supplies and supplies in informal settlements represent particular challenges. Loss of pressure during interruptions to supply exacerbates the impacts of other faults, such as backflow through cross-connections and ingress through faults and breaks in distribution systems. Informal settlements are commonly overcrowded, with poor sanitation; leakage is high, governance is typically lacking and maintenance is poor.

The health outcomes from distribution system–related outbreaks can be severe, but risks are preventable, providing sufficient attention is paid to preventing contamination. Effective implementation of the World Health Organization's (WHO) *Guidelines for Drinking-water Quality* (WHO, 2011) requires application of an integrated risk management framework based on a multiple-barrier approach that extends from catchment to consumer. This includes protection of water sources, proper selection and operation of treatment processes and management of distribution systems. In recent years, much attention has been paid to preventing contamination of water sources as a first step, followed by selecting and reliably operating treatment processes. Evidence from the USA shows that this has been successful in reducing waterborne outbreaks associated with inadequate treatment, particularly of surface water supplies (Craun et al., 2006; Craun, 2012). Despite this progress, the regular occurrence of outbreaks associated with distribution systems suggests that too little attention is being paid to sound management of these systems (Craun & Calderon, 2001; Craun et al., 2006). Contamination of distribution systems occurs after treatment, meaning that, with the exception of residual disinfectants, which provide some protection against bacterial and viral hazards, there are no further control measures for contaminants that gain entry to distribution systems or are released from pipe materials. Hazards introduced through faults in distribution systems typically flow directly to consumers. Integrity of distribution systems represents the final barrier before delivery of drinking-water to consumers, and management of risks in these systems should be incorporated in well designed water safety plans (WSPs).

The issue of maintaining water safety in distribution systems was identified as a significant concern at a meeting of experts convened by WHO in July 2011 in Singapore. This was based on existing evidence that inadequate design, construction and management contribute to a significant proportion of drinking water–borne disease. In addition, stresses caused by rapid urbanization, population growth and ageing infrastructure could further exacerbate problems with distribution systems. It was agreed that additional guidance on the application of good management practices for distribution systems was required.

The guidance provided in this document focuses on applying the framework for safe drinking-water (see Fig. 2), including WSPs, as described in the fourth edition of the *Guidelines for Drinking-water Quality* (WHO, 2011). The scope of this document includes small to large piped water systems in both developed and developing countries. It applies from the outlet of primary treatment processes to delivery to consumers, including at standpipes, but does not include pipework within buildings either before or after the point of delivery. This is the subject of the complementary text on *Water Safety in Buildings* (Cunliffe et al., 2011).

This document builds on earlier guidance provided in the supporting document on *Safe Piped Water: Managing Microbial Water Quality in Piped Distribution Systems* (Ainsworth, 2004) as well as other texts, including:
- *Health Aspects of Plumbing* (WHO & WPC, 2006);
- *Heterotrophic Plate Counts and Drinking-water Safety* (Bartram et al., 2003);
- *Pathogenic Mycobacteria in Water: A Guide to Public Health Consequences, Monitoring and Management* (Pedley et al., 2004);
- *Water Safety Plan Manual* (Bartram et al., 2009); and
- *Water Safety Planning for Small Community Water Supplies: Step-by-Step Risk Management Guidance for Drinking-water Supplies in Small Communities* (WHO, 2012).

This reference document is intended to be used as a supplementary document to help water suppliers and regulators who are already familiar with the WSP approach, as detailed in the *Water Safety Plan Manual* (Bartram et al., 2009) (i.e. water suppliers who require more specific technical assistance in developing and implementing the WSP approach in their distribution systems and regulators who support or audit WSP implementation). This tool should be applicable to both low- and high-resource systems.

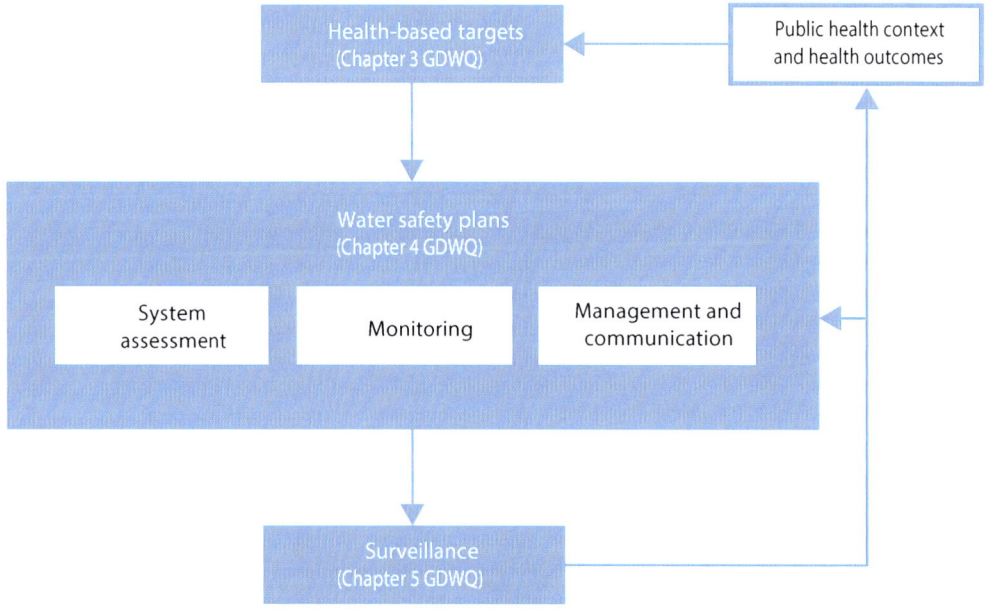

Fig. 2. The framework for safe drinking-water, as detailed in the WHO *Guidelines for Drinking-water Quality* (GDWQ)

The main text is divided into 12 sections following the descriptions in the *Guidelines for Drinking-water Quality* (WHO, 2011) and based on the 11 modules included in the *Water Safety Plan Manual* (Bartram et al., 2009), with an additional section describing the enabling environment (policy and regulations, independent surveillance and disease surveillance) (Fig. 3). It is important for regulatory and policy frameworks to support the implementation of WSPs to ensure their successful application. A number of case-studies are provided as annexes to illustrate the challenges that can confront drinking-water suppliers and potential solutions to overcome these challenges.

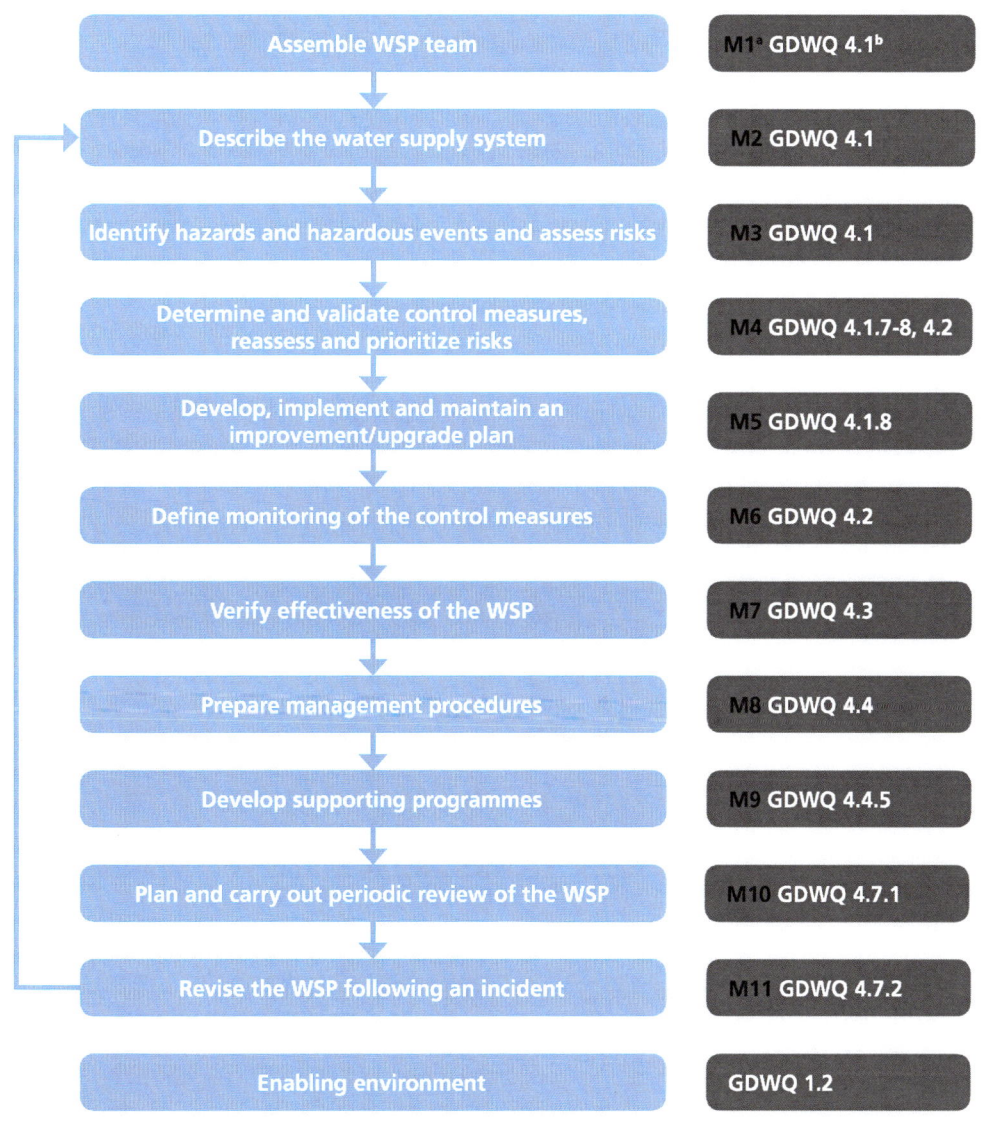

Fig. 3. Water safety plan steps and enabling environment. M1, M2, etc. indicate the relevant module from the *Water Safety Plan Manual* (Bartram et al., 2009); GDWQ 4.1, 4.2, etc. indicate the relevant chapter from the *Guidelines for Drinking-water Quality* (WHO, 2011).

Water safety plans

The most effective means of consistently ensuring the safety of drinking-water supplies is through the use of a comprehensive risk assessment and risk management approach incorporated in a WSP that applies to all steps of a water supply, including the distribution system. Normal practice is to develop an integrated WSP applying to all components, from catchment through treatment and distribution. However, in this document, the guidance focuses on application to distribution systems only. The aim of WSPs is straightforward: to consistently ensure the safety and acceptability of drinking-water supplies. In the case of distribution systems, it is assumed that water is safe to drink at the point of entry, so the aim becomes to maintain safety by preventing contamination after treatment. In simple terms, this includes:

- constructing systems with materials that will not leach hazardous chemicals into the drinking-water;
- maintaining integrity to prevent the entry of external contaminants;
- maintaining the supply of drinking-water to consumers; and
- maintaining conditions to minimize the growth of microbial pathogens (e.g. *Legionella*) and biofilms, scaling and accumulation of sediments.

Guidance on how to achieve and verify these outcomes is presented in the form of modules identified in the *Water Safety Plan Manual* (Bartram et al., 2009). Each of the modules from the *Water Safety Plan Manual* is discussed in more detail in the following sections.

1. Assemble the WSP team

- **Assemble WSP team M1**
- Describe the water supply system M2
- Identify hazards & hazardous events, assess risks M3
- Determine and validate control measures, reassess and prioritize the risks M4
- Develop, implement and maintain improvement plan M5
- Define monitoring of control measures M6
- Verify the effectiveness of the WSP M7
- Prepare management procedures M8
- Develop supporting programmes M9
- Plan and carry out periodic review of the WSP M10
- Revise the WSP following an incident M11

ENABLING ENVIRONMENT

- Engage senior management, and secure financial and resource support
- Identify the required expertise and appropriate size of the team
- Appoint a team leader
- Define and record the roles and responsibilities of the individuals on the team
- Define the time frame to develop the WSP

Assembling a qualified, dedicated team is a core preparatory step for the development and implementation of a WSP. The team should have knowledge of the water distribution system from source to the point of water consumption and the types of safety hazards to be anticipated, as well as the authority to implement the necessary changes to ensure that safe water is produced and supplied. The team may also include relevant stakeholders, such as public health agencies, standard-setting bodies and training providers, who play an important role in the provision of safe drinking-water. The membership of the team should be periodically reviewed, with new or replacement members brought in if required.

The key actions for ensuring success at this step include:
- engaging senior management, and securing financial and resource support;
- identifying the required expertise and appropriate size of the team;
- appointing a team leader;
- defining and recording the roles and responsibilities of the individuals on the team; and
- defining the time frame to develop the WSP.

2. Describe the water supply system

The first task of the WSP team is to fully describe the water supply system in order to support the subsequent risk assessment process. Describing the water supply system involves the following steps:
- gathering information on the system;
- preparing a flow chart from source to consumer and including the elements described below;
- inspecting the system to verify that the flow chart is accurate; and
- identifying potential water quality problems.

The description of the distribution system provides the foundation for development of the WSP. It will assist the WSP team in identifying where the system is vulnerable to hazardous events, relevant types of hazards and control measures.

2.1 Types of water transmission or distribution system

Usually, treated water is conveyed to service reservoirs for distribution to consumers. In urban systems, a water transmission system may also be necessary to convey water from a treatment plant to a number of service reservoirs located at different convenient points in the city. In some cities, there may be a number of sources and water treatment plants supplying service reservoirs and water distribution systems. These distribution systems may be separate or linked.

Both water transmission systems and water distribution systems are networks of pipes. However, water transmission systems have a tree-like configuration, whereas water distribution systems usually have loops. Sometimes, supply of water from the clear water tank at the treatment plant to various service reservoirs is by gravity (Fig. 4(a)). Often, treated water is either pumped directly to various reservoirs (Fig. 4(b)) or pumped to a main balancing reservoir, which, in turn, supplies water to various service reservoirs by gravity (Fig. 4(c)). Such systems are termed *complete gravity*, *direct pumping* and *combined gravity and pumping systems*, respectively.

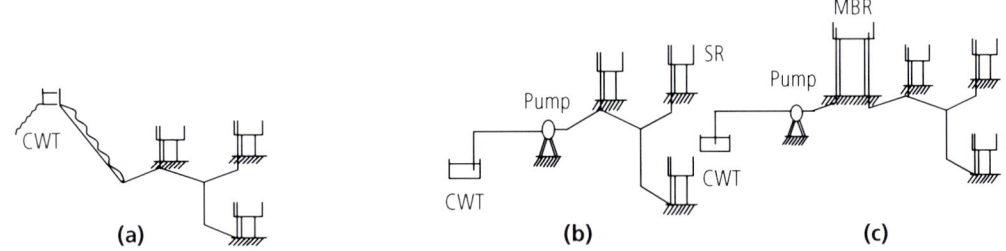

CWT: clear water tank; MBR: main balancing reservoir; SR: service reservoir

Fig. 4. Types of water transmission systems: (a) complete gravity; (b) direct pumping; and (c) combined gravity and pumping

The supply from the main balancing reservoir to various village/town reservoirs may be direct, as shown in Fig. 5(a). Such systems may be termed *single-level systems*. Sometimes, the main balancing reservoir may supply to several zonal balancing reservoirs, which, in turn, may supply several village reservoirs, as shown in Fig. 5(b). Such systems may be termed *multi-level systems*.

2. DESCRIBE THE WATER SUPPLY SYSTEM

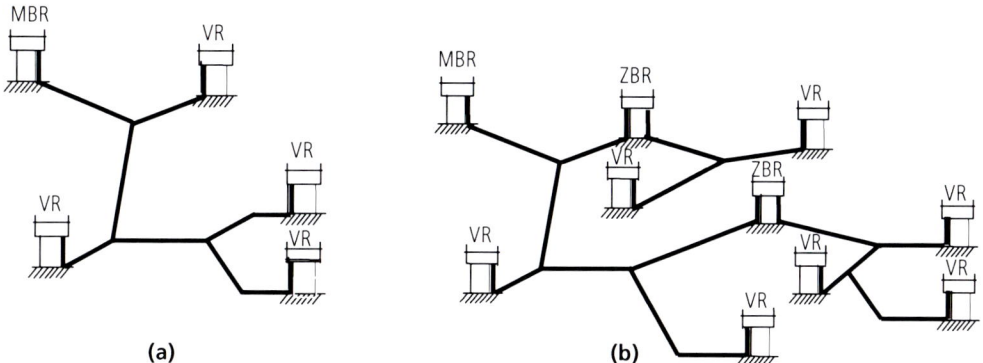

MBR: main balancing reservoir; VR: village reservoir; ZBR: zonal balancing reservoir
Fig. 5. Water transmission systems: (a) single level; (b) multi-level

2.2 Components of water transmission or distribution systems

Water distribution systems contain several components. Each network is unique in source, layout, topography of the service area, pipe material, valves and meters, and consumer connections. The description of the system should be as detailed as possible and include at least the location of major transmission and distribution mains, storages, secondary treatment devices (e.g. booster disinfection), pumps, valves and other fittings and standpipes (if in use).

A small distribution system may have a single source node, such as an elevated service reservoir or pumping arrangement directly supplying water from the sump (underground storage) at the treatment plant to the network. A large network, however, may have several source nodes – service and balancing reservoirs and pumping stations.

The layout of a distribution network depends on the existing pattern of streets and highways, existing and planned subdivision of the service area, property rights-of-way, possible sites for ground and elevated service reservoirs, and location and density of demand centres.

The topography of the service area may be flat or uneven. In an uneven terrain, booster pumps may be necessary for pumping water to high areas within the network. Similarly, it may be necessary to provide pressure-reducing valves for areas with lower elevation to reduce pressure. Check valves may also be necessary to maintain flow in the selected direction and restrict flow from the opposite direction.

Pipes in the distribution network may be of cast or ductile iron, mild steel, concrete and prestressed concrete, asbestos cement, polyvinyl chloride (PVC) and high-density polyethylene. Pipes may be unlined or lined with cement mortar.

Valves are provided in distribution systems to control flow, isolate pipelines during repairs and replacement, drain pipelines during cleaning, reduce pressure for low-lying areas, maintain flow in selected directions, suppress water hammer effects, allow air to enter pipelines while emptying, and release air at higher points and during filling. Meters are provided to measure flows of water from the source, transfer of water between zones and the supply of water to consumers. Hydrants with single or multiple outlets are provided in the network to supply water for fire extinguishing purposes.

Water supply to the tapping points of consumers may be direct or indirect. For direct supply, sufficient pressure is required in the network so that tapping points at different elevations get water. For multi-storeyed and high-rise buildings, however, the supply is indirect – the distribution system supplies water to sumps at the ground level, whereas lifting of water from these sumps to the individual tapping points and overhead tanks is left to the consumer. The supply to the consumer is controlled usually by a ferrule on the main, which is throttled sufficiently to deliver the required quantity of water at planned pressures.

2.3 Other important information and relevant performance indicators

Apart from the description of the water supply system, water quality information and performance indicators can be collected to assist the WSP team in carrying out risk assessments of vulnerabilities in the distribution system. For example, the quality of treated water after leaving the treatment plant can deteriorate as water travels through the pipe network. Some of the factors that influence changes in water quality between the treatment plant and the consumer's tap include:

- chemical and biological quality of the source water;
- effectiveness and efficiency of treatment processes;
- integrity of the treatment plant, storage and disinfection facilities and the distribution system;
- age, type, design and maintenance of the distribution network;
- time taken by the water to travel from the source to consumers' taps and presence of dead ends;
- water pressure;
- quality of treated water; and
- mixing of water from different sources within a distribution network and other hydraulic conditions (Clark, 1995).

Control of water quality is required right from the source to consumer points, and deterioration of water quality can be avoided in both water transmission and water distribution systems. Useful information that can be collected includes the following:

- information about pipe material, size and year of installation;
- pipe breakage history and leakage data;
- operational parameters, such as intermittency of supply, metering data and non-revenue water;
- environmental parameters, such as workmanship, bedding conditions and traffic;
- customer complaints regarding pipe breaks, supply of poor quality water and contamination;
- information on water quality – at source, after treatment and at consumer locations – and water age;
- type of supply, per capita supply, extent of metering and extent of non-revenue water;
- map showing open drainage and sewer networks;
- information about sewerage and drainage systems;
- solid waste dumping points; and
- information about illegal pumping by consumers.

The physical parameters (e.g. pipe age, material, diameter), operational parameters (e.g. intermittency of supply, number of breaks and bursts, leakage in the system) and environmental parameters (e.g. workmanship, bedding conditions, traffic) help to determine the vulnerability of pipes to contaminant intrusion. The zones of contamination formed in the soil near open drains and sewer crossings form a hazard. The risk of contamination intrusion can be assessed as a function of vulnerability and hazard.

The type of supply (continuous or intermittent), water pressure, total non-revenue water, per capita supply and extent of metering could be considered as performance indicators for water supply systems.

Finally, water age, defined as the time taken for the water to travel from source to consumer, is a factor influencing water quality deterioration within the distribution system. Water age is a hydraulic parameter and depends primarily on water demand, system operation and system design. Water age is significantly affected by storages, excessive capacities in the distribution system to meet emergency requirements and low demands during the initial period of design. Hydraulic analysis for an extended period can be performed using models such as EPANET to obtain water age (Rossman, 2000).

Table 1 provides a checklist of characteristics that could be included or considered in describing a water distribution system. The checklist is indicative, and not all characteristics will apply to all systems. The list is also not exhaustive, and other factors may need to be considered in describing some systems.

2.4 Informal settlements

Informal settlements represent particular challenges, with high population densities, poor sanitation, high water loss and leakage rates, and typically poor understanding of the extent of distribution systems. The presence of informal settlements in distribution systems should be identified, together with all relevant information about:
- the extent of distribution systems;
- water availability and interruptions to supply;
- the number and location of standpipes and kiosks;
- extent of metering;
- water leakage and non-revenue water; and
- extent of sewerage and drainage systems and alternative sanitation systems.

Table 1. Checklist of characteristics of distribution systems

Component	Characteristics to be considered
Physical components	Water transmission mains Pipe materials Pumps Treatment processes (e.g. supplementary chlorination) Metering Standpipes and water kiosks Water distribution mains Service reservoirs (storage tanks) Valves Hydrants Age of infrastructure
Water quality	Chemical and biological quality of source water Mixing "Dirty water" (black water, rust, turbidity), customer complaints Types of source water Water quality data Changes in distribution systems (e.g. disinfection by-products) Penetration of residual disinfectant
Physical factors influencing water quality	Cross-connection control programmes Frequency of water main breaks Soil conditions that could influence water main breaks Leakage (non-revenue water) History of external corrosion
Hydraulic (performance) factors influencing water quality	Water flows Intermittent or continuous supply Water pressure (including variations) Detention times (water age)
Environmental factors	Terrain Drainage systems Solid waste dumps Sewerage systems (proximity, open or closed systems)

3. Identify hazards and hazardous events and assess the risks

Hazardous agents, including microbial pathogens and chemical contaminants, that get access to drinking-water and distribution systems could affect the quality of the water and have an adverse impact on human health. This section describes the different microbial, chemical and physical hazards that can affect the safety of drinking-water in distribution systems. It also identifies the various hazardous events that could affect the water quality and the physical and hydraulic integrity of the distribution system, leading to water in distribution systems becoming contaminated or to supply being interrupted. This section also includes the process by which to assess potential risks to human health once the hazards and hazardous events are identified.

> **Definitions of hazards, hazardous events and risk**
>
> The *Water Safety Plan Manual* (Bartram et al., 2009) and the WHO *Guidelines for Drinking-water Quality* (WHO, 2011) define hazards, hazardous events and risk as follows:
>
> - A **hazard** is a biological, chemical, physical or radiological agent that has the potential to cause harm.
> - A **hazardous event** is an incident or situation that can lead to the presence of a hazard (what can happen and how).
> - **Risk** is the likelihood of identified hazards causing harm in exposed populations in a specified time frame, including the magnitude of that harm and/or the consequences.

3.1 Hazards

The following sections describe the major microbial, chemical and physical hazards in distribution systems, as summarized in Table 2.

3.1.1 *Microbial hazards*

Most microorganisms present in drinking-water distribution systems are harmless. However, infectious microorganisms may enter the distribution system and can survive and in some cases grow in the distribution system, increasing the potential for waterborne disease outbreaks (see Table 2).

Faecal contamination is a common source of infectious microorganisms. These include bacteria, viruses and parasites that occur naturally in the gut of humans and other warm-blooded animals. Faecal contamination of distribution system water may occur when a pathway – such as broken mains, intrusion, cross-connections or openings in storage tanks – has been established for faecal contaminant entry (USEPA, 2006). In addition, construction, new pipe installation and repairs close to sewer lines can introduce contamination into the distribution system. The presence of faecal pathogens is assessed by monitoring for indicator bacteria. The WHO *Guidelines for Drinking-water Quality* (WHO, 2011) recognize *E. coli* as the indicator of choice, although thermotolerant coliforms can be used as an alternative.

Microorganisms that grow in the environment may enter the drinking-water distribution system and attach to and grow on drinking-water pipes and other surfaces, forming biofilms, particularly at the ends of distribution systems where flows can be low (Power & Nagy, 1999; National Research Council, 2006; USEPA, 2006). Biofilms contain many sorption areas that can bind and accumulate organic and inorganic contaminants, as well as particulate and colloidal matter. These substances, also known as extracellular polymeric substances, protect microorganisms from biological, physical, chemical and environmental stresses, including predation, desiccation, flux changes and disinfectants (USEPA, 2006; Wingender & Flemming, 2011). For example, although viruses and protozoan cysts or oocysts need a

Table 2. Microbial, chemical and physical hazards that can be found in finished drinking-water, pipe biofilms and distribution systems.[a]

		Microbial hazards[b]	
Bacteria	**Viruses**	**Parasites**	**Filamentous fungi and yeast**
Campylobacter jejuni/ C. coli, Escherichia coli (some strains)	Noroviruses	*Cryptosporidium hominis/ parvum*	*Aspergillus flavus*
	Rotaviruses		*Stachybotrys chartarum*
Vibrio cholerae	Enteroviruses	*Entamoeba histolytica*	*Pseudallescheria boydi*
Salmonella typhi	Adenoviruses	*Giardia intestinalis*	*Mucor*
Other *Salmonella* spp.	Hepatitis A	*Cyclospora cayetanensis*	*Sporothrix*
Shigella	Hepatitis E	*Acanthamoeba*	*Cryptococcus*
Legionella spp.	Sappoviruses	*Naegleria fowleri*	
Non-tuberculous *Mycobacterium* spp.		Some invertebrates, including water mites, cladocerans and copepods	
Franciscella tularensis			

Chemical hazards[c]

Aluminium, antimony, arsenic, barium, benzo(a)pyrene, cadmium, chromium, copper, cyanide, disinfection by-products (including trihalomethanes, haloacetic acids and *N*-nitrosodimethylamine), fluoride, iron, lead, mercury, nickel, pesticides, petroleum hydrocarbons, selenium, silver, styrene, tin, uranium, vinyl chloride

Physical hazards

Turbidity, offensive odours, iron, colour, corrosion scales, sediment resuspension

[a] This table does not represent an exhaustive list of hazards in drinking-water, pipe biofilm or distribution systems.
[b] For further information on the human health effects, sources, occurrence, routes of transmission and significance of drinking-water as a source of infection for the microbial pathogens listed here, refer to Chapter 11 of the WHO *Guidelines for Drinking-water Quality* (WHO, 2011).
[c] For further information on chemicals and their primary sources, see Chapters 8 and 12 of the WHO *Guidelines for Drinking-water Quality* (WHO, 2011).

suitable warm-blooded host in which to grow, incorporation into biofilms could prolong their survival. Biofilms may develop within pipes and on other components, such as outlets, mixing valves and washers. Biofilms are extremely difficult to remove from all parts of the system once they are established, and they can be resistant to disinfectants. Hence, the goal is to control biofilm growth and not to eliminate biofilms totally.

The great majority of biofilm organisms are non-pathogenic, but they can include free-living and opportunistic pathogens (National Research Council, 2006; WHO, 2011) (see Table 2). Although most free-living heterotrophic bacteria, fungi, protozoa, nematodes and crustaceans in biofilms are not likely to be of risk to the drinking-water consumers, their activities can produce taste and odour problems, increase disinfectant demand and contribute to corrosion (National Research Council, 2006). Free-living pathogens linked to water-related diseases of most concern include *Pseudomonas aeruginosa*, *Legionella*, *Naegleria* and *Mycobacterium* spp. (National Research Council, 2005; WHO, 2011; Wingender & Flemming, 2011; CDC, 2014). Of these, *Legionella* is of increasing concern. For example, the cases of Legionnaires' disease in the USA have tripled in the past decade from 1110 in 2000 to 3522 in 2009 (CDC, 2011). The incidence rate increased from 0.39 to 1.15 per 100 000 people during that time. The increase may be due to a rise in the number of seniors and other people at high risk for infection. Amplification of *Legionella* concentrations within building plumbing systems is a major contributor to increased risk, but distribution systems can represent the source of the original inoculum (WHO, 2011).

The free-living pathogen *Naegleria fowleri* can also survive and multiply in biofilms, particularly in distribution systems that are susceptible to warming. *Naegleria fowleri* has caused deaths in Australia, Pakistan and the USA following nasal exposure to drinking-water containing the organism (Dorsch, Cameron & Robinson, 1983; Shakoor et al., 2011; CDC, 2014).

Microorganisms growing in biofilms can subsequently detach from pipe walls into the water under various circumstances, such as changes in the flow rate, and represent a potential source of water contamination.

3.1.2 Chemical hazards

There are many chemical hazards that could contaminate water in distribution systems. Some of the common chemical contaminants are unwanted residues of chemicals used in water treatment, chemicals coming from materials or reacting with materials in the distribution systems, chemicals accumulated in and then released from scales and deposits, and chemicals entering the distribution system through faults and breaks (Table 2).

3.1.2.1 Disinfection by-products

Disinfection by-products (DBPs) are produced by the reaction between chemical disinfectants and naturally occurring organic material in the source water (Krasner, 2009; WHO, 2011; Krasner et al., 2012). Guideline values have been established for a number of DBPs (WHO, 2011), and keeping concentrations below these values is encouraged. However, this should never compromise disinfection. Risks associated with inadequate disinfection are well established and far greater than potential risks from long-term exposure to DBPs.

Although there is a wide range of DBPs, trihalomethanes (THMs) and five haloacetic acids (HAAs) are generally considered to be good indicators of chlorination by-products (WHO, 2011). However, other DBPs, such as bromate, chlorate or *N*-nitrosodimethylamine (NDMA), may need to be considered when other types of disinfectant, such as ozone, chlorine dioxide and chloramines, are used (WHO, 2011). Recently, halobenzoquinones have been identified as potential bladder carcinogens (Du et al., 2013), whereas iodinated DBPs are of emerging interest (Krasner, 2009). This highlights that other novel DBP hazards not well indexed by THMs or HAAs may also be significant and that there is a need to stay current with research findings.

3.1.2.2 Chemicals from pipe materials and fittings

Materials that are in contact with drinking-water in public water distribution systems may leach agents into the water that are hazards to public health. Pipes, fittings, lubricants, o-rings, pipe and storage tank coatings, and mechanical devices have demonstrated the potential to leach contaminants with health concerns. There was evidence that some pre-1977 PVC piping products leached significant amounts of vinyl chloride into water (Flournoy et al., 1999). The most significant contributor of lead to drinking-water in many countries remains the presence of lead service lines in the distribution system. Lead may leach into potable water from lead pipes in old water mains, lead service lines, lead in pipe jointing compounds and soldered joints, lead in brass and bronze plumbing fittings, and lead in goosenecks, valve parts or gaskets used in water treatment plants or distribution mains. Lead was a common component of distribution systems for many years. The increased monitoring of lead in drinking-water starting in the 1980s demonstrated that lead leaching from brass devices could also be a source of lead (Lytle & Schock, 1996). The need to further reduce exposure to lead prompted new requirements for ultra low lead brass materials. NDMA has also been reported as coming from materials installed in the water distribution system (Morran et al., 2009, 2011).

While the full extent of the potential hazards of vinyl chloride and lead in water distribution systems may not have been realized until their occurrence was well documented, many national and international standards and accompanying certification programmes have been developed to proactively control this problem. These approval systems establish methods for prescreening products for their ability to add contaminants to drinking-water. This provides a mechanism for prevention of drinking-water contamination from distribution system materials.

There are several standards and approval systems across Europe that are working to create a harmonized process (e.g. Umweltbundesamt, 2012), as well as examples in North America (NSF International, 2012), Asia (Japanese Standards Association, 2012) and Australia (Standards Australia, 2005). These standards address both aesthetic effects, such as taste and odour (British Standards Institution, 2000; Standards Australia, 2005), as well as potential health effects (NSF International, 2012). Some standards address the potential of materials to support microbial growth (British Standards Institution, 2000; Standards Australia, 2005; DVGW, 2011). Standards address the health effects of chemicals that may be imparted to water from the materials in several different ways. Some standards set prescriptive requirements for the presence or amount of certain chemicals in materials. Other standards require toxicity tests on water that has been exposed to the material samples. Still other standards require measurement of the concentrations of contaminants imparted into water, which are then compared with established maximum threshold concentrations.

Approval systems or certification schemes typically include laboratory testing of products to the requirements of the standard. In addition, the schemes may require sampling of the product from the marketplace or surveillance of the production methods in the factory (ISO/IEC, 2013; NSF International, 2013b).

Since their introduction in North America in the 1920s, cement mortar linings have been used on the interior of cast iron pipe to provide a smooth protective barrier to the effects of corrosion. Over the last 90 years, the composition of mortars and methods of application have been improved to provide many years of reliable service. Cement mortar linings can, however, be susceptible to accelerated degradation under some water chemistry conditions.

Chemicals in drinking-water may also react with distribution system materials and their scales, causing the release of contaminants. Changing the disinfection treatment chemical from chlorine to chloramines was one of the factors that resulted in large releases of lead from the pipe material due to nitrification in the distribution system in Washington, DC, USA, in 2004 (USEPA, 2007a).

3.1.2.3 Water treatment chemicals

Water treatment chemicals are used to improve the quality of drinking-water; however, they can also be hazards if dosed at too high a concentration. One of the most unfortunate situations occurred with an accidental addition of aluminium sulfate into the water in Camelford, United Kingdom, in 1988, which resulted in 20 000 individuals being exposed to 3000 times the United Kingdom's permissible level of aluminium in drinking-water (Rowland et al., 1990).

Water treatment chemicals may contain trace levels of hazardous contaminants. Several national and international standards have been developed to limit the levels of contaminants in water treatment chemicals (Drew & Frangor, 2003). Over 200 European norms for water treatment chemicals have established maximum levels of contaminants that products are allowed to contribute to drinking-water (European Committee for Standardization, 2014). Other examples include NSF Standard 60 (NSF International, 2013a), used in both the USA and Canada, and NBR 15784 in Brazil (ABNT, 2009).

The WHO *Guidelines for Drinking-water Quality* identify chemical and material supplier certification programmes as an essential aspect of supporting programmes that should be documented in a WSP (WHO, 2011).

3.1.3 Physical hazards

Physical hazards refer to contamination affecting physical properties of water, such as colour, odour and turbidity.

Drinking-water that suddenly changes colour may indicate a hazard that should be addressed. Consumer complaints of green or blue water are typically associated with copper contamination from plumbing supplies. New copper installations and the use of excess flux during soldering are sometimes associated with the problem. Copper solubility is increased at low pH as well as high alkalinity, and corrosion control chemicals such as orthophosphate can decrease the leaching of copper (Schock, Lytle & Clement, 1995).

Brown, red and orange water complaints are usually the result of corrosion of iron or steel pipes. Changes in water quality or flow velocity in the distribution system are sometimes associated with the complaints (Benjamin, Sontheimer & Leroy, 1996).

Sediment in distribution systems that results in particles that settle to the bottom of a glass in a few moments may or may not be associated with colour effects. Sediments can range from harmless pieces of sand to scale containing heavy metals; in certain circumstances, sediment can indicate a breakage in the water main line, which could result in serious microbial or chemical contamination.

Taste and odour reports from consumers may indicate potential hazards associated with water distribution systems and should be investigated; however, they typically do not relate to actual hazards. While some taste and odour thresholds for chemicals are significantly lower than their health effects thresholds, the USEPA (2002b) reported that for many chemicals the organoleptic thresholds were significantly higher than the hazard-based regulatory limits.

Potentially serious hazards include petroleum contamination of drinking-water. While rare, there have been instances of permeation of drinking-water distribution materials by petroleum fuels (Glaza & Park, 1992). Except for benzene, most of these chemicals are detectable by taste and odour at concentrations that are well below health-related guideline values.

When sulfur, rotten egg, mouldy or musty odours or odours similar to grass, fish or earth emanate from the drinking-water, it may be a sign of bacterial contamination or bacterial growth.

3.2 Hazardous events

Hazardous events are incidents or situations that can lead to the presence of a hazard. In this section, hazardous events are grouped into three categories based on the circumstances or events that affect the integrity of the distribution system and the quality of the water within. These categories are:
- *Physical integrity*: breaks in the physical barrier of the distribution system that allow external contamination affecting the quality of the drinking-water, including structural failures of the distribution system components (pipes, valves, storage reservoirs), cross-connections, backflow and human activity (unsanitary activities during construction or vandalism);
- *Hydraulic integrity*: factors that could cause a water distribution system to lose its hydraulic integrity,

such as changes in flow and pressure caused by poor operational controls of valves and pumps and impacts of repairs and maintenance;
- *Water quality integrity*: situations that could cause a loss of water quality due to processes that take place within the distribution system, such as biofilm growth, leaching, corrosion, water age, stagnation/high retention times (due to dead ends) and discoloration.

Table 3 summarizes the typical hazardous events within a distribution system and is not intended to be exhaustive. The types of hazards involved are also indicated and identified as microbiological (M), physical (P) or chemical (C).

Table 3. Hazardous events associated with distribution systems

Category of event	Hazardous event	Hazard
System construction and repair		
Physical integrity	Contamination during construction of new water mains: • microbial or chemical contamination during construction or renovation due to debris, vermin, soil, groundwater or rainwater entering an open pipe (not capped) or fitting while the pipe/fitting is on the truck, stacked in the store yard, lying beside the trench or in the trench before connection	M, P, C
	Contamination of distribution system during new installations, including water meters, pumps, valve or hydrant insertions	M, P, C
	Contamination during water main repair: • an open main (not capped) when in the repair trench; could allow contamination, including petroleum products, from pumps used for dewatering • debris, soil or groundwater remaining in the main after repairs and not removed during the main recharge operation	M, P, C
Hydraulic integrity	Sediment resuspension, sloughing of biofilms causing customer complaints due to incorrect valve operation (closed or opened) after repairs	M, P, C
Water quality integrity	Contamination from impurities in materials used in construction and maintenance of pipes, fittings and tanks (e.g. copper, iron, lead, plasticizers, bituminous lining)	C
	The use of inappropriate materials, including use of metallic products that are incompatible with existing materials in the system, causing corrosion	C
System operation		
Physical integrity	Corrosion leading to loss of structural integrity	M, P, C
Hydraulic integrity	Contamination from leaky water mains in areas of low pressure or intermittent water supply: ingress due to backflow through leaky joints, air valves, perforations	M, P, C
	Contamination from leaky sewer mains in areas of low pressure or with intermittent water supply: ingress due to backflow through leaky joints, air valves, perforations, leaking valves and hydrants	M, P, C
	Accumulation of biofilms, sediments and particles in water mains due to low flow velocities in pipes and resuspension during high-flow events	M, P, C
	Resuspension of biofilms, sediments, scales due to flow reversals	M, P, C

Table 3 (continued)

Category of event	Hazardous event	Hazard
Water quality integrity	Discoloured water due to internal corrosion of unlined water mains (mild steel, cast iron, ductile iron) and accumulation of particles (e.g. sediments, manganese deposits), particularly at dead ends, due to long stagnation	P, C
	Survival of pathogens, growth of opportunistic pathogens and nuisance organisms in biofilms	M
	Elevated DBPs due to high levels of organic matter in source water	C
Storage tanks		
Physical integrity	Microbial contamination from entry of birds and small animals or faeces through faults and gaps in: • roofs or hatches • overflow pipes and inlet control valves from upstream sources • air vents	M
	Ingress of contaminated groundwater from unsealed joints and cracks	M, P, C
	Internal corrosion of steel water storage tanks	C
	Security breaches from unauthorized access by humans, including vandalism, sabotage	M, P, C
Water quality integrity	pH increases in concrete tanks due to excessive detention times	P
	Corrosion of internal fittings and surfaces	C
	Sediment accumulation and biofilm growth in the bottom of the tank	M, C
Backflow		
Physical integrity	Backflow from residential/industrial/commercial customers due to lack of prevention device or failure of device; likelihood increased during low-pressure events in water supply network	M, P, C
	Accidental cross-connection between drinking-water and non-drinking-water assets during construction or maintenance, including opening a normally shut valve to allow recharging after repairs and failing to close after completion	M, P, C
Secondary disinfection		
Water quality integrity	Excessive chlorine above health-based guideline value (5 mg/L)	C
	Underdosing of chlorine leading to inadequate protection against ingress of microbial contamination or growth of biofilms	M
	Elevated DBPs due to high levels of organic matter in source water	C

DBPs: disinfection by-products

3.2.1 Physical integrity
3.2.1.1 Physical faults, illegal connections, programmed shutdowns and recharge

Physical integrity refers to the maintenance of a physical barrier against external contamination. Failures that cause losses in physical integrity increase the risk of contamination. A breach in the system caused by corrosion or fracture can allow the ingress of contaminated groundwater or wastewater containing pathogens or harmful chemicals. The most common system failures are attributed to ageing and deteriorating piping system components, unstable ground conditions and sudden and excessive pressure fluctuations. Failures allow contaminant ingress when the system is under atmospheric or negative pressure conditions, as is common in intermittent water supplies, during a water main break or while the system is under construction or repair.

3. IDENTIFY HAZARDS AND HAZARDOUS EVENTS AND ASSESS THE RISKS

An illegal connection to the water supply can have a detrimental impact on the quality of water. An illegal water connection is where an individual tampers with, taps or makes a connection with a water supply without prior permission from the water utility. Connections made by unqualified persons without competency or knowledge of the safe design, installation and maintenance of the water supply system can expose it to unprotected cross-connections that can introduce microbial and chemical contaminants. Illegal connections also result in unaccounted for water loss that reduces system water pressure and increases the risk of contamination, as noted above.

3.2.1.2 Cross-connections and backflow

Cross-connection means any actual or potential connection or structural arrangement between a drinking-water system and any other source or system through which it is possible to introduce into the distribution system contaminated water, industrial fluid, gas or substances other than the intended drinking-water with which the system is supplied. Cross-connections constitute a serious public health risk. There are numerous well documented cases of cross-connections that contaminated drinking-water and resulted in serious illness (USEPA, 2002a; Craun et al., 2006).

Cross-connection with sewage system causing a *Campylobacter/Salmonella* outbreak in Nokia, Finland

In 2007–2008, approximately 5000 cases of gastrointestinal illness attributed to *Campylobacter* and *Salmonella* were reported in a population of about 30 000 in Nokia, Finland.

The outbreak was caused by cross-connection of the sewage system and the drinking-water system. Drinking-water was supplied through a permanent pipe at the sewage treatment plant for cleaning purposes. Contamination occurred when a worker undertaking repairs opened a valve connecting the two systems, and, due to a pressure differential, sewage flowed into the drinking-water system. The fault was not discovered for 2 days, in which time an estimated 450 000 litres of sewage entered the drinking-water system. The first signals of the contamination were customer complaints of odd colour, taste and smell, which were ignored. The reporting of cases led to recognition of the problem. High levels of faecal indicators were detected in the system, followed later by *Campylobacter*, *Salmonella*, viruses and *Giardia*. Hyperchlorination and flushing were used to decontaminate the system. The Finnish Defence Forces assisted with water carting and delivery of bottled water during the boil water advisory, which remained in place for 10 weeks. The outbreak and remedial action cost an estimated 3.7 million euros.

Source: Co-operative Research Centre for Water Quality and Treatment (2007a,b)

Contaminants enter the drinking-water system when the pressure of the contaminant source exceeds the pressure of the water system. This is known as backflow. The lower the system pressure and/or the increased instances of leakage in the piping network, the greater the probability of contaminant ingress.

In addition to physical faults in the distribution system, the backflow of contaminants can come from connections to non-potable systems, tanks, receptors, equipment or plumbing fixtures where inadequate cross-connection controls, including backflow prevention devices, have been installed or where maintenance has been inadequate (see sections 4 and 8). There are many examples of situations where devices were not maintained, where they were bypassed or where they did not perform effectively. Owners and managers of buildings can underestimate the potential impact of inadequate cross-connection controls on drinking-water distribution systems.

Although backflows through cross-connections have caused a broad and varied range of outbreaks of illness associated with drinking-water, surveys of water utilities have found that many do not have inspection programmes or have programmes that are insufficient to provide protection against cross-connections (USEPA, 2002a).

3.2.1.3 Contamination of storage reservoirs

As described in section 2, water supply systems broadly consist of storage reservoirs that can be used for storage and for balancing of water supply, transmission and distribution systems, and localized pressure booster systems.

Storage reservoirs in the water transmission or water distribution system are critical assets, as they can store large volumes of drinking-water, which could go beyond a day's supply for a large customer base. Reservoirs in the transmission system or at customers' premises or buildings may also perform the key role of regulating variations in water demand, as well as catering for supply and demand emergencies, such as power failure, firefighting and breakdown of the transmission system or supply from the treatment plant. They offer regulation, continuity and reliability of supply. In addition, elevated storage reservoirs ensure that pipelines have a net positive pressure, which prevents the ingress of groundwater, even in the event of disruptions during pumping from the treatment plant.

With the critical role that storage reservoirs play in water supply, it is important that best practices are incorporated into their design and construction to prevent microbial or chemical contamination of drinking-water from ingress of environment pollutants or faecal contamination from animals, birds and insects. Important design factors include water supply pressure, residence time of water, temperature, disinfectant residual, flow pattern, stratification and mixing of water to avoid stagnation, negative pressure conditions, backflow and outlet pipe position to maximize use of storage (Ainsworth, 2004).

Storage tank contamination leading to a *Salmonella* outbreak in Alamosa, Colorado, USA

In 2008, 434 cases of illness, including 20 hospitalizations and one death, were reported from a population of 8900 in Alamosa, Colorado, USA. Epidemiological investigations estimated that up to 1300 people could have become ill. The outbreak was caused by *Salmonella* Typhimurium. Alamosa was supplied with undisinfected groundwater from deep bores that had been consistently clear of faecal contamination.

In response to the outbreak of illness, it was decided to clean, flush and chlorinate the drinking-water system. This started at a roofed ground-level reservoir, which was drained to enable removal of sediment and disinfection. Workers noticed cracks and holes in the tank wall, and these faults were considered the likely entry point of contamination. *Salmonella* was detected throughout the distribution system fed from this tank. A snowmelt prior to the outbreak could have carried the *Salmonella* into the tank. The tank had not been subject to routine cleaning and inspections.

Disinfection was added to the supply after the outbreak. The cost of the outbreak was estimated at US$ 2.6 million.

Source: Colorado Department of Public Health and Environment (2009); Ailes et al. (2013)

Some examples of possible hazardous events that may pose a risk to drinking-water quality in storage reservoirs include intrusion of sediments, small animals or insects through faults such as:
- damage to roofs, including gaps in hatches and covers;
- gaps between the roof structure and the tank wall;
- cracks in concrete tank walls or corrosion of metal tanks;
- gaps at entry points of pipework or cables; and
- splits in membrane liners.

In addition, intrusion of physical, chemical or biological contaminants into storage reservoirs could also be a result of sabotage or mischief by intruders with motivation for contaminating the water.

As water from storage reservoirs is typically supplied to a large customer base, contamination of the treated water in the storage reservoir, whether intentional or accidental, could result in significant public health consequences.

3.2.1.4 Plumbing issues and protecting the water supply at the customer interface

Protecting the quality of water within the main water distribution system is the responsibility of the water utility. However, once water from the main water distribution system passes onto the property of an individual user, it is the owner of the property along with a competent plumber who must protect the water supply on the premises. Although water safety in buildings is generally not within the scope of this publication, faults within building systems can have impacts on water quality in distribution systems through backflow.

Drinking-water supply piping, water outlets and equipment within and around buildings need to be protected from all contaminated or polluted liquids or substances at all times and in all instances. Cross-connections with sources of contaminants, such as chemical storages and non-potable water supplies, need to be prevented.

There are two types of cross-connections – direct and indirect. There are two types of backflow caused by these cross-connections – back-siphonage and back-pressure backflow.

In a direct cross-connection, there must be a direct physical connection existing between the drinking-water and contaminant source. A greater pressure must also exist on the downstream side of this connection, causing a reversal of flow from the original intended direction of flow. Backflow occurs when the pressure within a polluted system exceeds the pressure in the drinking-water system. This is a back-pressure cross-connection. A typical example of this type of cross-connection is the supply line to a non-potable system using a pump to circulate the fluid. The water in the non-potable system and connected piping represents the hazard. This is because the pump circulating the fluid in the piping system could cause an increase of pressure above the supply pressure and cause a backflow of non-potable water into the drinking-water supply.

There are two types of indirect cross-connections: under-rim or submerged connections, and over-rim connections. The under-rim or submerged connection is where the drinking-water inlet comes into the bottom or side of a receptacle and is immersed in a polluted or contaminated substance (see D in Fig. 6(a)). Without some form of protection, just the filling of the fixture to its rim could cause backflow into the drinking-water supply.

An over-rim connection is one where the water supply terminates above the flood-level rim of a fixture but has a hose fitting or connection that creates a potential for an under-rim termination (see A, B and F in Fig. 6(a)). In this instance, the over-rim supply line may not be continuously submerged unless a hose is permanently attached.

Backflow may occur if the hose is left in the sink and something happens to cause a negative or lower pressure in the fixture water supply line, such as a water main break or draining the system while leaving the faucet or tap open. This will cause a siphon to occur, drawing the possibly contaminated water into the drinking-water supply. The type of backflow that occurs in these instances is termed back-siphonage. See Fig. 6(a) for other examples of back-siphonage.

The control of backflow requires the removal of one of the two essential factors that can cause the backflow – namely, the physical link or the cause of the negative or low pressure. Removal of the physical link or cross-connection, such as in Fig. 6(b), is a positive means of preventing backflow and is the only true means of preventing contamination. This can also be accomplished with a backflow prevention device or an air gap. The appropriate selection, installation and testing of the devices are functions essential to the process of isolating pollutants and contaminants from the drinking-water system.

There are two basic locations where backflow protection is installed. The first is at the connection of the public water supply to the water service on the property of an individual user. This is known as containment backflow protection, because the public water supply is contained and protected from contamination that may occur on the property of an individual user.

The other location where backflow protection is installed is at the intersection of the drinking-water supply and the potential sources of contamination. This is known as isolation backflow prevention because it isolates each potential contaminant source from the drinking-water supply within or around a building. An atmospheric vacuum breaker installed on the water supply to a urinal is an example of isolation protection.

3. IDENTIFY HAZARDS AND HAZARDOUS EVENTS AND ASSESS THE RISKS

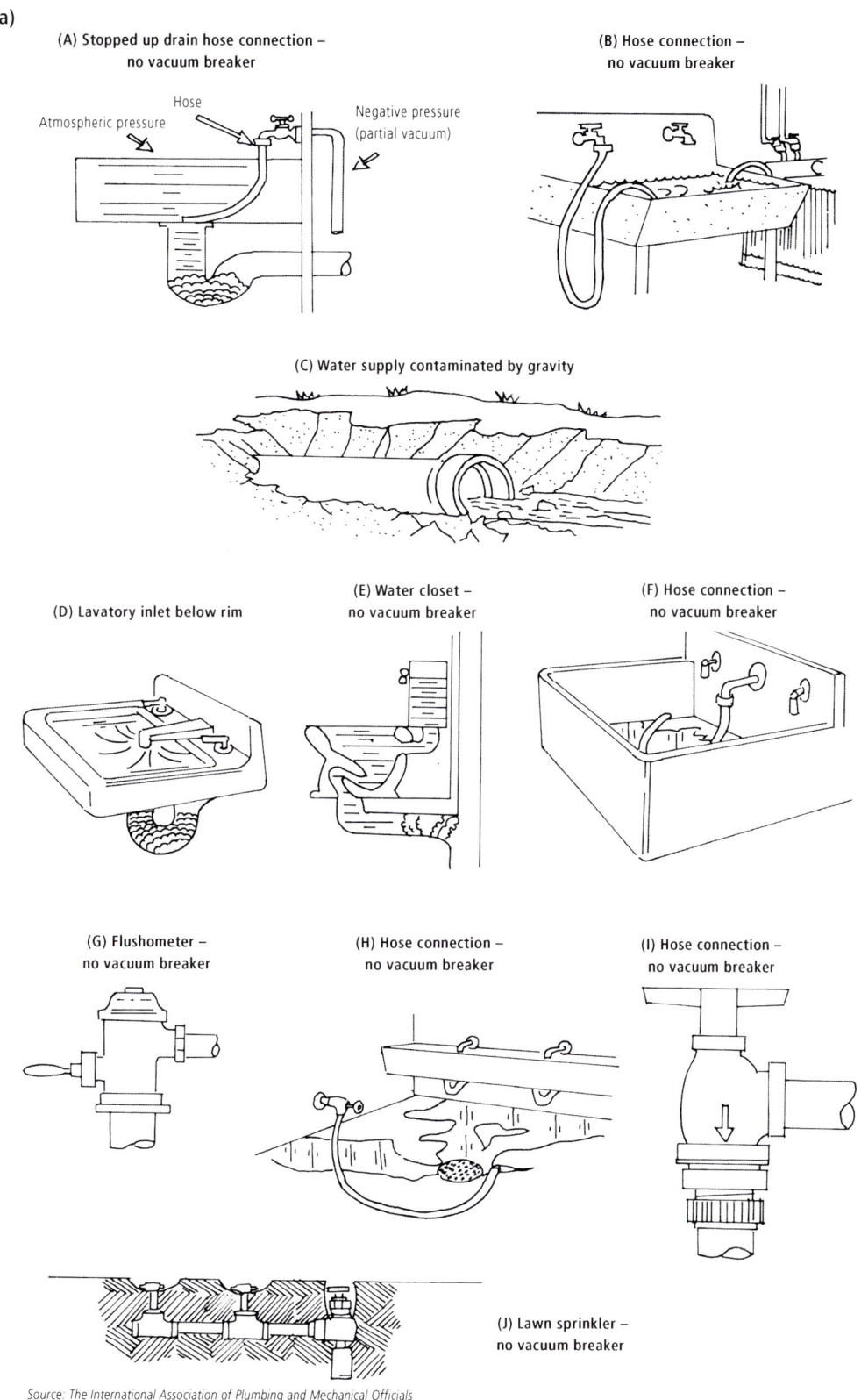

Fig. 6. (a) Cross-connections and other sources of contamination; (b) direct link cross-connection; (c) separation of water supply and sewer piping; (d) air gap

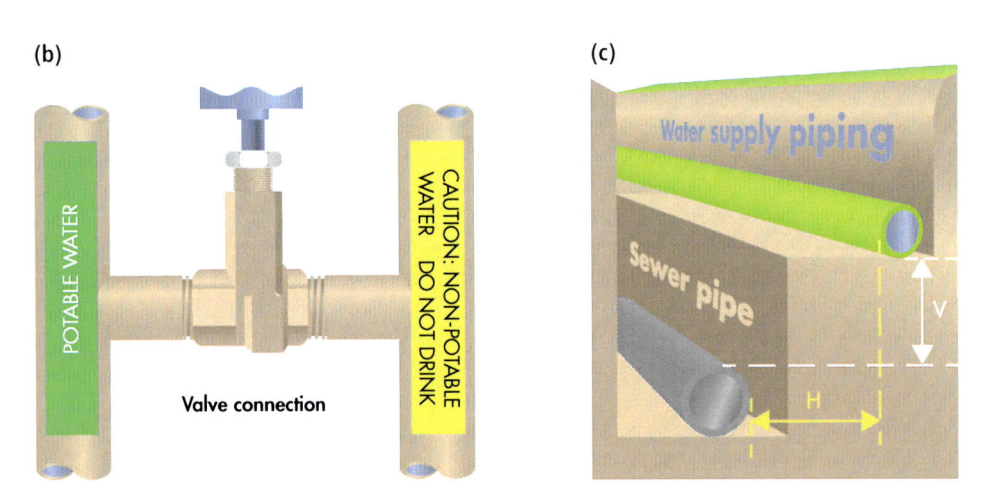

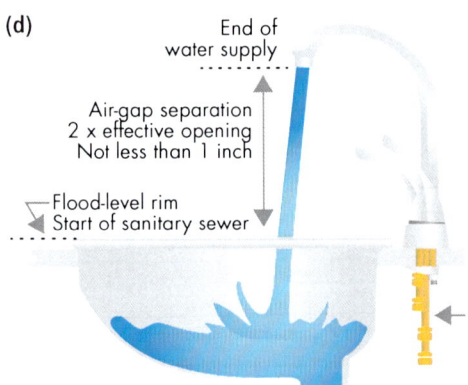

Fig. 6 (continued). (a) Cross-connections and other sources of contamination; (b) direct link cross-connection; (c) separation of water supply and sewer piping; (d) air gap

3.2.1.5 Construction work, renovations and repairs

The distribution system must provide a secure barrier to contamination; it should be fully enclosed and kept under positive pressure to deliver water to user taps and prevent contamination entering the water main.

The significant benefit (i.e. barrier) of positive pressure is highlighted by the example of a small hole developing in the water main; the water will leak out of the main because the water is under positive pressure or at a higher pressure than the water in the surrounding soil. The soil and any potential contaminants in the soil will not enter the main while the water is leaking out under pressure – that is, the hydraulic gradient is from the main to the soil.

However, the positive pressure in the water main may drop from time to time by accident or under control by the water utility during construction, renovation or repair works. It is at this controlled time when flow is stopped, when the water in the main is drained away and when the main is cut open for a period that potential ingress of contamination can occur. This is the critical time for water utilities. If not properly designed, planned and managed, these construction/repair works can lead to the introduction of microbial and chemical hazards.

Water main breaks and poor repairs associated with an *E. coli* O157 outbreak in Cabool, Missouri, USA

In 1989–1990, 243 cases of illness attributed to *E. coli* O157 were reported in a population of 2100 people in Cabool, Missouri, USA. There were 82 cases with bloody diarrhoea; 32 cases were hospitalized, two developed haemolytic uraemic syndrome and two died. Cabool was supplied with undisinfected groundwater from deep bores. Monitoring data from the previous 10 years confirmed that the bores were free from faecal contamination.

The town water supply was implicated as a cause of the outbreak. The distribution system was poorly maintained, with 35% of the water flow unaccounted for, suggesting leaks and poor metering. The sewerage system was also in poor condition, operating above capacity with regular spills. In December 1989, Cabool experienced extremely low temperatures, which caused two large water mains and 45 in-ground water meters to fail. Most cases occurred after the second water main break and after replacement of the water meters. Sewage overflow through the water main breaks was suspected as a primary source of the outbreak, but this was not confirmed by water testing. Only two water samples were analysed, and neither was from the area with the highest incidence of cases. Poorly managed repairs of the mains and replacement of the meters were suspected to be contributing factors. There was limited flushing of mains after repairs, there was no disinfection and there was no collection of water samples to verify microbial quality after completion of the repairs.

Chlorination of the water supply was introduced in January 1990.

Source: Hrudey & Hrudey (2004); National Research Council (2006)

A detailed description of common hazardous events that could occur during any construction, renovation or repair works on a distribution system is provided in Table 3.

3.2.2 Hydraulic integrity

The hydraulic integrity of a water distribution system represents the capacity to provide reliable quantities of water at acceptable pressures. Many events and faults that cause loss of physical integrity can cause loss of hydraulic integrity. Similarly, loss of positive pressure due to poor hydraulic integrity can worsen impacts of breaches in physical integrity, such as water main breaks and cross-connections. Loss of hydraulic integrity can also have impacts on water quality by increasing water age and promoting biofilm development. Components that can influence water pressure include pumps, pressure-reducing valves, non-return valves, hydrants and storage tanks. Events that can reduce pressure include pump failure, flushing, service tank cleaning, construction and renovation, water main breaks and repairs, and firefighting.

3.2.2.1 Changes in water flow due to intermittent water supply

The most extreme examples of lost hydraulic integrity occur in systems that fail to deliver an uninterrupted supply of drinking-water. Intermittent water supplies present serious health risks, in addition to being extremely inconvenient and in many cases wasteful. Intermittent water supplies are generally used when the available supply or the hydraulic capacities of the system are incapable of meeting demand. Intermittent water supplies are very susceptible to contamination due to:
- piping systems subjected to atmospheric and vacuum (negative pressure) conditions after supply hours, which allow the ingress of contaminated groundwater or wastewater through leaky pipes and joints;

- pipe corrosion and failure caused by exposing the system alternately to water and to air;
- periods of stagnation that promote microbial growth and make centralized disinfection difficult and ineffective;
- fluctuating pressures and flow velocities that cause biofilm detachment and release of microorganisms; and
- microbial growth in water stored in roof tanks for use between supply events.

Low water pressure and disease transmission in Uzbekistan and Tajikistan, 1996–1997

Uzbekistan

An epidemiological survey in Uzbekistan in 1996 showed that households with piped water had a much higher rate of diarrhoeal disease (75.5 cases per 1000 people per month) than households with no piped water that chlorinated drinking-water used in their home (28.8 cases per 1000 people per month). Source water entering the piped system was subject to two-stage chlorination and was not considered the cause of disease. However, 30% of households connected to the piped system received drinking-water containing no detectable chlorine residual, and 42% of households reported intermittent water pressures. It was concluded that the likely source of disease transmission was cross-contamination between the piped water system and leaky sewage pipes or contaminated groundwater due to low water pressure and incomplete protection of the water distribution system by chlorine residual.

Tajikistan

Between 1 January and 30 June 1997, 8901 cases of typhoid fever, including 95 deaths, were reported in Dushanbe, Tajikistan. Faecal coliform contamination was detected in 97% of tap water samples tested. Poor treatment contributed to disease transmission, but distribution system contamination was also significant. Low and intermittent water pressure and water outages were common. Some water pipes were run inside stormwater drains, and connections to pipelines were poorly controlled.

Source: Semenza et al. (1998); Mermin et al. (1999)

3.2.2.2 Changes in water flow due to mixing of water sources

Some drinking-water supplies can receive water from a number of sources, including different reservoirs, combinations of groundwater and surface water, and inputs from desalination plants or highly treated sewage (direct and indirect potable reuse). Contributions from individual sources can be varied for operational reasons, to meet regulatory requirements or in response to climatic conditions. For example, inputs from desalination plants or groundwater supplies may be increased when reservoir water levels are low. Similarly, water levels can lead to changes in inputs from different reservoirs.

Mixing of water supplies can cause a number of issues, including:
- changes in flow direction, leading to dislodging of biofilms and resuspension of sediments;
- changes in chemical quality, dislodging deposits (e.g. mixing desalinated water with hard surface water or groundwater);

- loss of disinfectant residuals when water with higher demand is mixed with water with lower demand or when incompatible disinfectants (e.g. free chlorine and chloramines) are mixed;
- reduced effectiveness of disinfectants due to pH and temperature changes;
- changes in taste; and
- impacts on specialist users (industrial, agriculture, etc.) due to changes in chemical composition.

3.2.3 Water quality integrity

Water quality integrity refers to maintaining quality by minimizing impacts caused by internal processes and events during delivery of drinking-water through transmission and distribution systems.

3.2.3.1 Release of hazards from materials and fittings, including corrosion and scaling

Hazardous chemicals can be released from materials and equipment by leaching due to contact with drinking-water or by corrosion. Leaching can be reduced by selecting materials and fittings that are suitable for contact with drinking-water, whereas corrosion can be reduced by operational controls applied by water utilities. Leaching tends to start immediately after installation, whereas corrosion increases with the age of the materials and fittings.

A number of countries have established certification programmes for materials and fittings used in contact with drinking-water, which include testing for the release of hazardous chemicals (section 3.1.2). New installations should use only materials and fittings that comply with these requirements. In some countries, specific materials, such as lead pipes and solders, have been banned due to their propensity to leach hazards.

However, these materials still exist in older installations pre-dating certification and bans. In addition, in the absence of certification programmes, material control may be poor; in some cases, despite the existence of certification programmes, material control may still be insufficient. For example, PVC piping used in drinking-water systems should not be manufactured using lead-based lubricants and stabilizing materials, but this same restriction does not apply to PVC piping used for irrigation. Selection of the wrong type of piping can result in leaching of lead into drinking-water (Mangas & Fitzgerald, 2003).

Hazardous chemicals that can be leached from materials and fittings used in distribution systems are summarized in Table 4. Of the chemicals listed in Table 4, lead has received the most attention on public health grounds due to its solubility in water and low guideline value (10μg/L). Copper corrosion is also relatively common due to the high level of use of copper pipes within buildings. Evidence of copper corrosion is often manifested in visible blue-green discoloration of bathroom and toilet fittings.

In areas where iron pipes are used, "red water" caused by the release of iron oxides is also a frequent source of customer complaints. Although iron and iron oxides do not represent a direct health concern, iron hydroxides can adsorb and concentrate chemicals such as arsenic. An investigation of iron pipe scales from 15 utilities in the USA found that arsenic was present at high concentrations even in systems where source water contained arsenic at concentrations below 10μg/L (Lytle, Sorg & Frietch, 2004). Arsenic accumulated in scales can be resuspended or released into drinking-water supplied to consumers following changes in water flow or disinfection. In the Midwest of the USA, the introduction of chlorination to a water system supplying groundwater led to the release of arsenic from iron deposits at concentrations approaching 5 mg/L (Reiber & Dostal, 2000).

Table 4. Common hazardous chemicals that leach from pipe materials and fittings

Material	Hazard
Lead pipes	Lead
Copper pipes	Copper
Solders and fittings (brass and bronze alloys)	Lead, copper, cadmium, nickel, silver, tin
Iron pipes	Iron, arsenic
Galvanized pipes	Zinc, lead
Cement pipes and tanks and cement mortar	Calcium hydroxide (high pH)
Polyethylene pipes, liners, jointing and sealing compounds	Organic compounds
Rubber seals	NDMA (from reaction with chlorine/chloramines)

NDMA: *N*-nitrosodimethylamine

The range of factors that can influence the release of chemicals from materials and fittings includes:
- age of materials and fittings;
- water age/stagnation;
- pH;
- alkalinity;
- temperature;
- chlorine/chloramine residuals;
- chloride and sulfate;
- aggressive water (soft water, low alkalinity); and
- changes in treatment.

However, it can be difficult to identify independent impacts of individual factors. For example, assessing the impacts of chlorine and chloramine residuals can be confounded by consideration of alkalinity and pH. In some cases, elevated release of hazardous chemicals can occur when factors coincide – for example, when water stagnation coincides with ageing materials or low-pH water (Health Canada, 2009).

Age of materials and fittings

The age of materials can either increase or decrease the release of hazardous chemicals, depending on the mechanism of release – that is, leaching or corrosion. Leaching of chemicals tends to be greatest during the first weeks and months following installation, whereas concentrations of chemicals released by corrosion generally increase with age. Lead is a good example of this differential response. The release of lead from newly installed brass alloys and solders occurs relatively rapidly over the first few weeks and months and then declines (Boffardi, 1988; Schock & Neff, 1988), whereas lead released by corrosion from lead pipes increases with age (Schock, Wagner & Oliphant, 1996). Organic chemicals tend to be released by leaching from newly installed materials, such as tank liners and jointing materials. In Australia, NDMA was produced in high concentrations during the first months after installation of rubber jointing materials due to the reaction of chloramines in the water supply with organic compounds in the jointing materials (Morran et al., 2009, 2011). These concentrations decreased over time. The use of polyelectrolyte coagulants (i.e. quaternary amine–based coagulation polymers) and ion exchange resins has also been shown to release NDMA (Wilczak et al., 2003).

The age of iron pipe and hydraulic disturbances influence the release of iron. Although scale buildup initially reduces corrosion, the eventual release of this scale from old tuberculated cast iron pipes can result in "red water" (Sarin et al., 2003).

Leaching of lime from concrete pipes, linings and mortars generally starts immediately after installation and decreases over time (Douglas & Merrill, 1991; Douglas, Merrill & Catlin, 1996; USEPA, 2002a). Leaching can also occur when cement-based materials deteriorate (LeRoy et al., 1996).

Water age/stagnation

Stagnation tends to exacerbate leaching and corrosion. It also increases concentrations of hazardous chemicals in water by reducing the impacts of dilution associated with water flows. Leaching of lead and copper from pipes and fittings and lime from cement can be increased by stagnation (Wong & Berrang, 1976; Douglas & Merrill, 1991; Schock, Lytle & Clement, 1995; Douglas, Merrill & Catlin, 1996; Sorg, Schock & Lytle, 1999; Lytle & Schock, 2000). In contrast, copper leaching tends to be curtailed by the low dissolved oxygen levels found in stagnant water (Sorg, Schock & Lytle, 1999; Lytle & Schock, 2000).

pH

The release of hazardous chemicals from materials and fittings can be influenced by pH. Generally, increasing pH values lead to decreased release of metals, largely due to decreased solubility at higher pHs.

Solubilization of lead from pipes and alloys and galvanic corrosion of lead from solders all decrease at higher pHs (Schock, 1989; Reiber, 1991; USEPA, 1992; Singley, 1994). Surveys of water utilities have found that pHs above 8–9 correlated with decreased lead at consumers' taps (Karalekas, Ryan & Taylor, 1983; Lee, Becker & Collins, 1989; Dodrill & Edwards, 1995; Health Canada, 2009). As a result, raising the pH in water supplies has been used as a control measure to reduce lead concentrations.

Similarly, results from over 350 water utilities showed that none exceeded the United States Environmental Protection Agency's (USEPA) action level for copper of 1.3 mg/L, provided that the water had a pH of 7.8 (Dodrill & Edwards, 1995). Copper corrosion occurred under two sets of conditions: pH below 7.0 and alkalinity below 30 mg/L (as calcium carbonate) and pH below 7.8 and alkalinity above 90 mg/L (as calcium carbonate). In both cases, corrosion was reduced by increasing the pH (Dodrill & Edwards, 1995).

As pH increases, corrosion of iron pipes can increase, but iron concentrations in drinking-water decrease due to the lower solubility of iron at higher pHs (Karalekas, Ryan & Taylor, 1983; Pisigan & Singley, 1987; Broo, Berghult & Hedberg, 2001; Sarin et al., 2003).

Alkalinity

Alkalinity can have varied impacts on the release of hazardous chemicals from materials and fittings. Higher alkalinities decrease corrosion and the release of iron from pipes (Pisigan & Singley, 1987; Cantor, Park & Vaiyavatjamai, 2000; Sarin et al., 2003) and lime from cement pipes (Conroy et al., 1994). In contrast, water utility and laboratory results show that higher alkalinities increase copper release (Schock, Lytle & Clement, 1995; Edwards, Schock & Meyer, 1996; Edwards, Jacobs & Dodrill, 1999; Cantor, Park & Vaiyavatjamai, 2000; Shi & Taylor, 2007).

The relationship between alkalinity and the release of lead is not clear. Two studies showed that lead release was reduced at higher alkalinities (Dodrill & Edwards, 1995; Cantor, Park & Vaiyavatjamai, 2000), but an earlier survey failed to identify a relationship between lead concentrations in drinking-water and alkalinity (Lee, Becker & Collins, 1989).

Chlorine/chloramine residuals

The clearest evidence of corrosion associated with oxidizing disinfectants such as chlorine and chloramines comes from studies in which the concentrations of disinfectants or the disinfectants themselves (e.g. from chlorine to chloramines) were changed. This is discussed in more detail in section 3.2.3.2.

Free chlorine residuals have been shown to increase copper release at pHs below 7.5 (Reiber, 1989; Cantor, Park & Vaiyavatjamai, 2000; Boulay & Edwards, 2001), but this copper release was greatly reduced or reversed at higher pHs (Edwards & Ferguson, 1993; Edwards, Jacobs & Dodrill, 1999; Boulay & Edwards, 2001). Similarly, iron corrosion has been reported to increase in the presence of free chlorine (Pisigan & Singley, 1987; Cantor, Park & Vaiyavatjamai, 2000), but increasing the pH above 7.8 decreased corrosion (Cantor, Park & Vaiyavatjamai, 2000).

Evidence from laboratory studies and utility surveys indicated no increase in lead corrosion associated with free chlorine residuals (Lee, Becker & Collins, 1989; Cantor, Park & Vaiyavatjamai, 2000).

There is little evidence on the impacts of chloramines on corrosion. One study found that corrosion was reduced in chloraminated water but concluded that this was probably due to an increase in pH used to stabilize the chloramine residuals (Thomas, 1990). Reiber (1993) found that chloramines accelerate the corrosion of copper at pH 6, but this is unlikely to occur in drinking-water systems, as higher pHs are required to stabilize chloramine residuals.

Temperature

There is conflicting evidence on the impacts of temperature on corrosion. A survey of water utilities found no significant correlation between temperature and lead or copper concentrations in drinking-water (Dodrill & Edwards, 1995). In contrast, seasonal variations have been reported, with higher lead and iron concentrations being detected in warmer months of the year (Karalekas, Ryan & Taylor, 1983; Horsley et al., 1998). It should be noted that temperature influences a range of parameters that may also have an impact on corrosion, including chlorine and chloramine residuals, biological activity and water flows.

Chloride and sulfate

Chloride and sulfate can influence lead release. A survey of water utilities demonstrated that those with a chloride to sulfate mass ratio of less than 0.58 had lower lead concentrations than those with ratios greater than 0.58 (Edwards, Jacobs & Dodrill, 1999). Edwards & Triantafyllidou (2007) demonstrated that elevated chloride to sulfate ratios increased lead release from lead tin solders used in drinking-water systems.

Aggressive water

Soft, low-alkalinity water has been associated with leaching, particularly from cement-based materials. Cement-based pipes and mortars will leach lime in the first months and years after installation (Douglas & Merrill, 1991; Douglas, Merrill & Catlin, 1996). The level of leaching depends on the total alkalinity and buffer capacity of the transported water, the type of cement used, the contact time between the water and the cement material, and the pipe diameter (Bonds, 2005; AWWA, 2008). Providing alkalinity and water flows are maintained at reasonable levels to dilute any leached lime, this will generally not cause problems. However, at low flows and low alkalinity, lime leaching can cause large pH increases, with values above 11 being recorded (LeRoy et al., 1996; Vik et al., 1996).

Leaching can also extend to compounds contained in cement linings and mortars. Illness and mortalities at a renal dialysis centre were attributed to the leaching of aluminium from cement mortar–lined ductile pipe caused by aggressive (negative Langelier Index), soft (15–20 mg/L hardness as calcium carbonate), low-alkalinity desalinated water. In the 2 months after installation of the pipe, aluminium concentrations in the drinking-water increased from 5 µg/L to 690 µg/L; over the next 2 years, concentrations above 100 µg/L were recorded (Berend & Trouwborst, 1999). Water passing through cement mortar pipe can also leach substantial amounts of barium, cadmium and chromium for the first 14–18 days of water stagnation and increase pH, alkalinity and calcium concentrations for up to 4 years (Douglas, Merrill & Catlin, 1996; Gou, Toomuluri & Eckert, 1998).

Changes in treatment

Impacts of changes in treatment processes are discussed in more detail in the next section, but changes such as modifying pH and coagulation and changing disinfectants can all have impacts on rates of corrosion in the distribution system. For example, changing from chlorination to chloramination in Washington, DC, USA, contributed to the increased release of lead into the drinking-water supply (USEPA, 2007a).

3.2.3.2 Treatment process changes with an impact on distribution

Changes in treatment processes, such as enhanced coagulation or replacement of chlorination with chloramination, can be implemented to improve water quality. However, they can also have unintended consequences, including increased corrosion (USEPA, 2007b).

Examples of possible treatment changes and their consequences are shown in Table 5.

Table 5. Advantages and disadvantages associated with changes in water treatment processes

Treatment change	Advantage	Disadvantage
Modified coagulation	• Reduced concentrations of precursors of DBPs • Improved removal of *Cryptosporidium* and *Giardia* • Improved disinfection (lower demand)	• Increased corrosion associated with lower pH, lower alkalinity, unless adjusted, increased chloride and sulfate • Increased inorganic chemicals
Replacement of chlorine with chloramines	• Increased residual in system providing protection against external contamination and growth of free-living pathogens • Biofilm control	• Increased corrosion • Initial release of biofilms from parts of the distribution system that previously did not receive water containing residual disinfectant • Nitrification • Potential issues for dialysis patients • Decreased initial disinfection requiring longer contact times • Production of NDMA • Mixing chlorinated and chloraminated supplies
Supplementary chlorination	• Increased residual in system providing protection against external contamination and growth of free-living pathogens • Biofilm control	• Increased corrosion • Initial release of biofilms • Increased DBPs
Decreased pH (in chlorinated systems)	• Enhanced primary disinfection by chlorine • Lower DBPs by decreasing chlorine dose	• Increased corrosion • Decreased residuals through distribution system

DBP: disinfection by-product; NDMA: *N*-nitrosodimethylamine
Source: Adapted from USEPA (2007b)

Modified coagulation

Modifications to coagulation can be relatively short-term actions applied in response to changes in source water characteristics to maintain drinking-water quality or longer-term enhancements to improve drinking-water quality. Modifications can include increasing or decreasing coagulant doses, changing coagulants, adding polymers, adjusting pH or improved mixing. Enhanced coagulation can increase the removal of natural organic matter, subsequently reducing the formation of DBPs (Krasner & Amy, 1995; USEPA, 2007b) and increasing disinfection effectiveness by decreasing disinfectant demand. Enhanced coagulation can also increase the removal of pathogens such as *Cryptosporidium* and *Giardia* (States et al., 2002).

Modifications to coagulation, including changes in coagulants, may require adjustments to be made to subsequent filtration and disinfection processes. The adjustments can be determined by carrying out jar testing and enhanced surveillance of operational monitoring parameters to ensure that turbidity and total organic carbon control is maintained and disinfection is not adversely affected.

Other disadvantages include increased corrosion and increased inorganic compounds associated with higher coagulant dose rates. Modifying coagulation can lead to reduced pH, changes in the chloride to sulfate ratio (depending on the choice of coagulant) and reduced alkalinity (USEPA, 1999, 2007b). Adjusting the pH after filtration and prior to entry of the treated water into the distribution system can greatly reduce impacts on corrosion.

Replacement of chlorine with chloramines as a primary or secondary disinfectant

Although chloramines are weaker disinfectants than chlorine, their much longer persistence can greatly extend the distribution of disinfectant residuals, particularly in long or complex systems. This can increase protection against recontamination of distribution systems from external contamination and reduce the growth of free-living pathogens such as *Naegleria fowleri* and *Legionella* (Christy & Robinson, 1984; Thomas, 1990; Kool, Carpenter & Fields, 1999; Flannery et al., 2006).

Other advantages include better control of biofilms and reduced production of total THMs (Kirmeyer et al., 2004; WHO, 2004). However, this can be offset by the increased production of other DBPs, such as NDMA (WHO, 2008). It can also increase the occurrence of non-tuberculous mycobacteria, which may include opportunistic pathogenic species (Revetta et al., 2013).

Disadvantages of replacing chlorination with chloramination are:
- potential increases in corrosion;
- nitrification and growth of opportunistic pathogens (e.g. non-tuberculous mycobacteria);
- short-term discoloration of water at consumers' taps due to the release of biofilms from parts of the distribution system that previously did not receive water containing residual disinfectant;
- issues for dialysis patients;
- initial disinfection; and
- issues with blending chloraminated and chlorinated water.

In a new system, chloramines should cause fewer problems with corrosion than chlorine, as they are weaker oxidizing agents and normally are used in conjunction with higher pHs to maintain stability. However, when chloramination is used to replace chlorination, the change can lead to the release of accumulated scale by increasing solubility. Replacing chlorine with chloramines was one factor that contributed to the release of lead in the Washington, DC, USA, water supply (USEPA, 2007a).

The introduction of chloramination can cause short-term problems with discoloration due to the inactivation and release of biofilm material from pipes that previously did not receive water containing a residual disinfectant.

Biological nitrification occurs predominantly in storage reservoirs, in tanks and at the ends of chloraminated distribution systems and is caused by the oxidation of ammonia to nitrite and then nitrate, principally by nitrifying bacteria (Cunliffe, 1991; USEPA, 2002d). Nitrification leads to loss of chloramine residuals, growth of biofilm organisms and decreased pH, alkalinity and dissolved oxygen concentrations (USEPA, 2002d). Nitrification can be promoted by excess free ammonia, warm temperatures and low chloramine residual (Wolfe et al., 1990; USEPA, 2002d). Optimizing dosing to minimize excess ammonia and maintaining chloramine residuals, particularly at the ends of distribution systems, can reduce the likelihood of nitrification.

Owing to their persistence and stability in drinking-water, chloramines have caused haemolytic anaemia, methaemoglobinaemia and haemolysis in dialysis patients (WHO, 2004). Authorities responsible for dialysis must be informed prior to the replacement of chlorination with chloramination so that they can adjust the treatment of water used in dialysis.

Blending chlorinated and chloraminated supplies can be a challenge, as the correct balance between ammonia and chlorine concentrations is essential for maintaining effective disinfection. One response can be to apply sufficient chlorine first to eliminate the chloramine residual (chlorine to ammonia ratio greater than 7.6 : 1, in milligrams per litre) and second to provide a free chlorine residual.

Supplementary chlorination

Supplementary chlorination partway through the distribution system can improve the maintenance of free chlorine residual and enhance microbial control. The disadvantages are the potential to increase concentrations of DBPs and increase corrosion caused by chlorine.

Decreased pH

Reduced pH can be used as a mechanism to improve the effectiveness of primary chlorination, leading to the use of lower dose rates and decreased THM formation. As discussed previously, decreased pH can lead to increased corrosion of metals from pipes and fittings. Decreased pH can also reduce the removal of manganese (USEPA, 2007b).

3.2.3.3 Microbial growth and biofilms

Several conditions may lead to the occurrence of microbes and the formation of biofilms in distribution systems, including (USEPA, 2006; National Research Council, 2006):
- source water with high dissolved or particulate organic matter or close to sources of faecal matter;
- poor water temperature control;
- changes in water flow and stagnation;
- neutral pH in drinking-water;
- microbial interactions with pathogens such as *Acanthamoeba*, *Hartmanella* and *Naegleria* (USEPA, 2002e; Declerck et al., 2009);
- low oxygen;
- piped distribution system and certain pipe materials;
- inadequate cleaning and maintenance of distribution systems;
- loss of disinfectant residual;
- water main failures and breaks; and
- conditions of storage facilities, such as high-volume tanks that support stagnation and stratification or uncovered storage facilities.

For further reading on biofilm prevention and residual disinfectant:

Costello JJ (1984). Postprecipitation in distribution systems. J Am Water Works Assoc. 76(11):46–9.

Geldreich EE, LeChevallier M (1999). Microbiological quality control in distribution systems. In: Letterman RD, editor. Water quality and treatment, fifth edition. New York (NY): McGraw-Hill; 18.1–18.49.

Kirmeyer GJ, Friedman M, Martel KD, Howe D, LeChevallier M, Abbaszadegan M, et al. (2001). Pathogen intrusion into the distribution system. Denver (CO): American Water Works Association Research Foundation.

Trussell RR (1999). Safeguarding distribution system integrity. J Am Water Works Assoc. 91(1):46–54.

United States Environmental Protection Agency (1992). Control of biofilm growth in drinking water distribution systems. Washington (DC): United States Environmental Protection Agency (EPA/625/R-92/001).

van der Kooij D (2000). The unified biofilm approach: a framework for addressing biological phenomena in distribution systems. International Distribution Research Symposium. Denver (CO): American Water Works Association.

van der Kooij D, van Lieverloo JHM, Schellart J, Hiemstra P (1999). Maintaining quality without a disinfectant residual. J Am Water Works Assoc. 91(1):55–64.

3.2.3.4 *Water ageing*

Water age has been reported as a major factor in the deterioration of water quality within distribution systems, leading to public health and aesthetic concerns (USEPA, 2002c). It was also discussed above in section 3.2.3.1 with regards to the release of hazards from materials and fittings. With respect to health impacts, with increased water age, there is an increased potential for the formation of DBPs, increased corrosion and an increased potential for microbial regrowth, recovery and shielding. Increased water age can also lead to a loss in the effectiveness of corrosion control measures as well as to an increased potential for nitrification of the bulk water. With respect to aesthetics, with increased water age, there can be an increase in water temperature and in issues associated with taste, odour and colour. The two main mechanisms for deterioration of water quality are interactions with the pipe wall and the water, and reactions within the bulk water itself.

Water age can vary significantly within a system and is primarily controlled by system design and system usage. Based on a survey of more than 800 utilities in the USA, the Water Industry Database (AWWA, 1992) indicates an average distribution system retention time of 1.3 days and a maximum retention time of 3.0 days, although the literature reports that water age can be significantly longer, especially in dead end areas (Acker & Kraska, 2001; USEPA, 2002c).

Factors contributing to increased water age include demand planning and the requirements for providing capacity to deal with events such as power outages and firefighting. Planning necessitates the installation of facilities that have excess capacity for water storage and distribution, preparing to meet demands that may occur 20 years into the future and more. Building future capacity can in the short term increase water age, as the storage capacity may be large in comparison with present-day needs.

Building in capacity for both reserves and the delivery of water for use in the case of fire can also increase water age, not only from the additional reserve required, but also from the larger pipe diameters needed to accommodate fire flow. For example, every mile (1.6 km) of 4-inch (10.2 cm) pipe that is replaced with 8-inch (20.3 cm) pipe increases the effective volume of the distribution system by more than 10 000 gallons (39 000 L) (Prentice, 2001).

Water age will also be affected by periods of reduced water usage. This includes seasonal variations in demand, as water use typically varies over the course of the year, with higher demands occurring in the warmer months (National Research Council, 2006). This also includes instances where dramatic changes occur following the relocation of a significant percentage of water system consumers, as happened following Hurricane Katrina. Census figures for the USA show that the population of New Orleans was 29% less in 2010 than it was in 2000 (Campbell, 2011).

Hazards, hazardous events and risk assessment in informal settlements

Hazards, hazardous events and risks are magnified in informal settlements compared with adjoining urban areas. Population densities are much higher, sanitation is typically poorer, with very limited sewerage systems, non-revenue water rates are high, and most residents receive water from water kiosks and standpipes. Management of water systems in informal settlements is often poor. Hence, the likelihood of hazardous events occurring is much greater.

Some of the higher risks include:

- poor sanitation due to:
  - open defecation and "flying toilets",
  - overflowing sewers, pit latrines, bio-toilets and storm drains,
  - open drainage and
  - poor solid waste management;
- inadequate supply due to:
  - lack of storages,
  - high leakage rates,
  - illegal water closures by competing gangs operating kiosks and
  - high cost of laying pipes;
- poor pipe materials;
- illegal connections;
- inadequate protection of standpipes;
- poor hygiene;
- poor reporting and communication of leaks and other failures;
- limited access, restricting maintenance and repairs;
- limited governance, planning and operation;
- low ownership of services, leading to theft of pipes and fittings.

Source: Nairobi City and Water Sewerage Company and Athi Water Services Board (2009); Macharia (2012)

3.3 Risk assessment

Once potential hazards and hazardous events have been identified, the levels of risk need to be assessed so that priorities for risk management can be established. Risk assessments need to consider the seriousness of hazards and hazardous events in the context of likelihood of occurrence and consequence of exposure.

3.3.1 Semi-quantitative risk assessment

The WHO *Guidelines for Drinking-water Quality* (WHO, 2011) and the *Water Safety Plan Manual* (Bartram et al., 2009) both recommend assessing risk using a matrix based on likelihood and severity of consequences, as shown in Table 6.

Table 6. Example of risk assessment matrix

Likelihood	Severity of consequences				
	Insignificant or no detectable impact **1**	Minor impact on compliance **2**	Moderate aesthetic impact **3**	Major regulatory impact[a] **4**	Catastrophic public health impact[b] **5**
Almost certain (once per day) **5**	5	10	15	20	25
Likely (once per week) **4**	4	8	12	16	20
Moderately likely (once per month) **3**	3	6	9	12	15
Unlikely (once per year) **2**	2	4	6	8	10
Rare (once every 5 years) **1**	1	2	3	4	5
Risk score		<6	6–9	10–15	>15
Risk rating		Low	Medium	High	Very high

[a] Regulatory impact means exceeding regulatory limits with potential public health impacts.
[b] Catastrophic public health impacts include potential for outbreaks of illness with severe outcomes.

Source: Adapted from Bartram et al. (2009); WHO (2011)

The descriptions shown in Table 6 can be adjusted, as can the number of categories. It is important, before commencing a risk assessment, for the WSP team to establish what it means by terms such as minor, moderate, major, etc. Although many hazards may threaten water quality, not all will represent a high risk. The aim should be to distinguish between high and low risks so that attention can be focused on mitigating high and very high risks that are more likely to cause harm, through application of control measures (see section 4).

3.3.2 Quantitative microbial risk assessment

Quantitative microbial risk assessment (QMRA) is a process by which the impact of events such as ingestion of, inhalation of or contact with pathogens in drinking-water can be assessed. QMRA takes into consideration the sources of pathogens, fate and transport within the distribution system and exposure pathways. While it is unlikely that QMRA will be used on a routine basis, it may be a valuable

3. IDENTIFY HAZARDS AND HAZARDOUS EVENTS AND ASSESS THE RISKS

tool to support implementation of water safety planning, improve understanding of vulnerabilities of drinking-water distribution systems, assess risks associated with hazardous events and support risk management.

QMRA has been used to assess risks from microbial contamination of distribution systems, with one study finding that enteroviruses could represent a high risk when soil or shallow groundwater contaminated distribution systems, whereas *Campylobacter* may be a high risk if contamination was caused by surface water or sewage (van Lieverloo, Blokker & Medema, 2007). A second investigation found that *Campylobacter* was a likely source of contamination from storage reservoirs (Westrell et al., 2003). Such investigations reinforce the need for preventive measures, such as those designed to minimize entry of soils and groundwater during maintenance activities or entry of contaminants into storage reservoirs.

The document *Quantitative Microbial Risk Assessment for Water Safety Management: A Harmonized Approach to the Implementation of QMRA in the Water-related Context* (WHO, in preparation) describes the steps of a QMRA and facilitates the understanding of these steps in the practical implementation of QMRA to support water safety management (Fig. 7).

1. Problem formulation
What is the scope and purpose of the assessment?

- Which hazards?
- Which hazard pathways?
- Which health outcomes?

2. Exposure assessment
What is the predicted dose of pathogens for the defined hazard pathways?

- Source concentration
- Barrier reduction and recontamination
- Exposur volume, population size and frequency

3. Health effects assessment
What are the expected health effects of the defined hazards?

- Dose–response models
- Illness and sequelae
- Secondary transmission and immunity
- Impact on disease burden

4. Risk characterization
What is the expected health effect of the predicted dose?

- Quantification of risk
- Variability and uncertainty analysis
- Sensitivity analysis

Fig. 7. Framework for water-related QMRA (WHO, in preparation)

4. Determine and validate control measures, reassess and prioritize the risks

- Assemble WSP team **M1**
- Describe the water supply system **M2**
- Identify hazards & hazardous events, assess risks **M3**
- **Determine and validate control measures, reassess and prioritize the risks M4**
  - Determine current control measures
  - Validation of control measures
  - Reassess and prioritize risks
- Develop, implement and maintain improvement plan **M5**
- Define monitoring of control measures **M6**
- Verify the effectiveness of the WSP **M7**
- Prepare management procedures **M8**
- Develop supporting programmes **M9**
- Plan and carry out periodic review of the WSP **M10**
- Revise the WSP following an incident **M11**

ENABLING ENVIRONMENT

4.1 Determine current control measures

Control measures are barriers necessary for preventing or reducing significant water quality risks. They need to be developed, implemented and monitored for each hazardous event identified as significant in the risk assessment. In the context of the distribution system components, control measures are defined as those measures required in drinking-water distribution systems that directly affect the safety or aesthetics of drinking-water, either by preventing the occurrence of hazards or by inactivating, removing or reducing them to acceptable levels.

Control measures can include a wide range of activities and processes. They can be:
- preventive (and incorporated in design, planning and construction processes and renewal of infrastructure);
- treatment related (e.g. secondary disinfection);
- technical (e.g. operational and maintenance procedures); and
- behavioural (e.g. customer awareness programmes).

Control measures must be defined specifically and precisely for each significant hazardous event and adapted to the local conditions. They should never be imprecise or vague. Whereas the type and number of control measures will vary for each supply system, their collective implementation and maintenance are essential to ensure that water quality is controlled effectively. Only the current control measures being implemented by the water supplier should be included in the risk assessment.

Table 7 provides a list of control measures for typical hazardous events in the distribution system. Some of the control measures are applied during the design and construction of the water distribution system, whereas others involve a range of operational, emergency and programmed operating procedures (e.g. water main break repair procedure, water main cleaning, tank maintenance and cleaning, secondary disinfection, customer complaint management and other routine operating procedures) (see section 8).

Table 7. Examples of control measures

Hazardous event	Control measure
System construction and repair	
Contamination during construction of new water mains: • microbial or chemical contamination during construction or renovation due to debris, vermin, soil, groundwater or rainwater entering an open pipe (not capped) or fitting while the pipe/fitting is on the truck, stacked in the store yard, lying beside the trench or in the trench before connection	• Construction standards and specifications (including materials – storage, handling, transport, flushing, swabbing, disinfection, contact time and water quality testing) • Field compliance audits
Contamination of distribution system during new installations, including water meters, pumps, valve or hydrant insertions	• Code of practice • Construction standards and procedures • Disinfection practices prior to commissioning

4. DETERMINE AND VALIDATE CONTROL MEASURES, REASSESS AND PRIORITIZE THE RISKS

Hazardous event	Control measure
Contamination during water main repair: • an open main (not capped) when in the repair trench; could allow contamination, including petroleum products, from pumps used for dewatering • debris, soil or groundwater remaining in the main after repairs and not removed during the main recharge operation	• Dewatering of trench prior to commencing repairs • Prevention of contamination of pipe material during storage, transport and repairs ***For mains ≤150 mm*** • Flushing of water main – specify duration based on length of the main and minimum flow rate ***For mains >150 mm*** • Pipe cleaning (swabbing) and disinfection after repair • Water quality testing (visual/turbidity) prior to turning on the water main
Sediment resuspension, sloughing of biofilms causing customer complaints due to incorrect valve operation (closed or opened) after repairs	• Standard operating procedures for operation of valves after repairs
Contamination from impurities in materials used in construction and maintenance of pipes, fittings and tanks (e.g. copper, iron, lead, plasticizers, bituminous lining)	• Approved product standards for materials in contact with water • Approved product list • Compliance audits and materials checklist • Replacement of lead service lines
The use of inappropriate materials, including use of metallic products that are incompatible with existing materials in the system, causing corrosion	• Approved product standards for materials in contact with water • Approved product list • Compliance audits and materials checklist

System operation

Hazardous event	Control measure
Corrosion leading to loss of structural integrity	• Approved product standards for materials in contact with water • Approved product list • Leak detection programme • Pipe and fittings replacement programme
Contamination from leaky water mains in areas of low pressure or intermittent water supply: ingress due to backflow through leaky joints, air valves, perforations	• Maintain positive pressure, provide continuous supply • Maintain minimum chlorine residuals in the distribution network; if necessary, install secondary/booster chlorination • Leak detection and repair programme • Pipe and fittings replacement programme • Design and construction specifications and standards
Contamination from leaky sewer mains in areas of low pressure or with intermittent water supply: ingress due to backflow through leaky joints, air valves, perforations, leaking valves and hydrants	• Maintain positive pressure, provide continuous supply • Maintain minimum chlorine residuals in the distribution network; if necessary, install secondary/booster chlorination • Leak detection and repair programme (sewer and water main) • Pipe and fittings replacement programme (sewer and water main) • Design and construction specifications and standards • Design and construction standards to maintain separation between water and sewer mains
Accumulation of biofilms, sediments and particles in water mains due to low flow velocities in pipes and resuspension during high-flow events	• Design standards to achieve self-cleaning pipe velocities • Operate valves and pumps to avoid rapid surges in flows • Routine water main cleaning programme (in areas where self-cleaning velocities cannot be achieved) • Maintain minimum chlorine residuals in the distribution network; if necessary, install secondary/booster chlorination

Table 7 (continued)

Hazardous event	Control measure
Resuspension of biofilms, sediments, scales due to flow reversals	• Operate valves and pumps to avoid flow reversals where possible • Routine water main cleaning programme • Maintain minimum chlorine residuals in the distribution network; if necessary, install secondary/booster chlorination
Discoloured water due to internal corrosion of unlined water mains (mild steel, cast iron, ductile iron) and accumulation of particles (e.g. sediments, manganese deposits), particularly at dead ends, due to long stagnation	• Routine water main cleaning programme • Water main condition and criticality assessment and inspection programmes to prioritize replacement programme • Water main renewal programme • Improving contaminant reduction at treatment plant (e.g. minimize manganese level well below aesthetic level)
Survival of pathogens, growth of opportunistic pathogens and nuisance organisms in biofilms	• Controls to prevent pathogen intrusion due to ineffective treatment or distribution system integrity breaches (e.g. treatment targets, main repair procedures) • Maintain minimum chlorine residuals in the distribution network; if necessary, install secondary/booster chlorination • Replacement of chlorination with chloramination • Reducing or preventing biofilm growth through proper maintenance
Elevated DBPs due to high levels of organic matter in source water	• Additional treatment to remove precursors – dissolved organic matter in source water (e.g. coagulation, magnetic ion exchange) • Alternative disinfection – chloramination • Reducing detention times (e.g. eliminating dead ends, increasing turnover through storage tanks during periods of low flow by dropping high water levels, taking tanks out of service during low flows)
Storage tanks	
Microbial contamination from entry of birds and small animals or faeces through faults and gaps in: • roofs or hatches • overflow pipes and inlet control valves from upstream sources • air vents	• Reservoir inspection and maintenance programme, including repair of faults/gaps • Disinfect tank after repairs • Maintain minimum chlorine residuals in the distribution network; if necessary, install secondary/booster chlorination • Design and construction standards
Ingress of contaminated groundwater from unsealed joints and cracks	• Reservoir inspection and maintenance programme • Design and construction standards
Internal corrosion of steel water storage tanks	• Cathodic protection • Routine inspection and maintenance programme
Security breaches from unauthorized access by humans, including vandalism, sabotage	• Security fencing, locked gates, access hatches, alarms, routine security patrols, closed-circuit television cameras
pH increases in concrete tanks due to excessive detention times	• Operate system to avoid excessive detention • Lower volume of water in tanks to increase turnover during periods of low flow
Corrosion of internal fittings and surfaces	• Design and construction standards • Reservoir inspection and maintenance programme
Sediment accumulation and biofilm growth in the bottom of the tank	• Reservoir cleaning programme included in routine inspection and maintenance programme

4. DETERMINE AND VALIDATE CONTROL MEASURES, REASSESS AND PRIORITIZE THE RISKS

Hazardous event	Control measure
Backflow	
Backflow from residential/industrial/commercial customers due to lack of prevention device or failure of device; likelihood increased during low-pressure events in water supply network	• Installation of appropriate backflow prevention device based on level of risk (low, medium and high) – e.g. water meters with built-in dual check valve for residential customers, testable backflow prevention device for high-risk industrial customers • Annual inspection programme (e.g. test reports, recording of devices in water supplier's database) • Backflow prevention code/standards • Maintain positive pressure in distribution system
Accidental cross-connection between drinking-water and non-drinking-water assets during construction or maintenance, including opening a normally shut valve to allow recharging after repairs and failing to close after completion	• Construction standards and specifications – specifically mentions measures for prevention of cross-connections, such as asset identification in plans/geographic information system • Training of operational staff • Adequate identification or marking on site and on plans
Secondary disinfection	
Excessive chlorine above health-based guideline value (5 mg/L)	• Monitor chlorine residuals and vary dose so that chlorine residuals stay within limits
Underdosing of chlorine leading to inadequate protection against ingress of microbial contamination or growth of biofilms	
Elevated DBPs due to high levels of organic matter in source water	• Additional treatment to remove precursors – dissolved organic matter in source water (e.g. coagulation, magnetic ion exchange) • Reducing detention times (e.g. eliminating dead ends, increasing turnover through storage tanks during periods of low flow by dropping high water levels, taking tanks out of service during low flows)

DBPs: disinfection by-products

A comprehensive set of benchmark security measures for water storage reservoirs and approaches for consequence mitigation is presented in the *Guidelines for the Physical Security of Water Utilities* (AWWA, 2004).

Whereas control measures are directed at ensuring water quality, there may also be preventive actions and responses applied to maintain constancy of supply. These could include long-term system upgrades to ensure continuous supply in areas of low pressure or intermittent supply.

Control measures in informal settlements

There is a range of control measures that can be applied to decrease risks and improve water quality. However, the challenges are significant, and implementation is likely to be demanding.

- Improved sanitation:
  - Sewered ablution blocks, improved pit latrines, construction of bio-toilets
  - Maintain sewers and storm drains to reduce overflows
  - Clean and maintain open drains
  - Community solid waste collection systems
- Improved supply:
  - Provision of storages/tanks
  - Increase number of kiosks
  - Coordinate purchase of piping and fixtures
- Audit systems, improve metering and meter reading, remove or formalize illegal connections
- Repair leaks from standpipes, ensure that the area around the standpipe is protected from sewage and ponded water, ensure that outlets are kept clean
- Public health education to improve hygiene
- Raise public awareness of impacts of faults, establish communication procedures, improve ownership of infrastructure to reduce theft
- Establish coordinated operation and maintenance programme
- Establish formal governance structures supported by legislation and regulations

The effectiveness of these control measures can be monitored by assessing:
- cleanliness of areas;
- incidence of disease;
- frequency of sewage and stormwater drain overflows;
- reduced leakage and non-revenue water, improvements in meter reading;
- number of water kiosks and water tanks; and
- public awareness.

Source: Nairobi City and Water Sewerage Company and Athi Water Services Board (2009); Macharia (2012)

4.2 Validation of control measures

All control measures should be validated to ensure their effectiveness. Validation is the process of obtaining evidence that control measures will be effective and achieve the required results. In other words, validation answers the question, "Will the control measures work?" Validation justifies the residual risk assessment scores assigned after consideration of the effectiveness of the control measures, as shown in Table 9 in the next section. Validation can take the form of:
- analysis of water suppliers' historical data (e.g. operational and verification water quality data and field audit analysis to confirm that the current water main repair practice is adequate to remove contamination after a water main repair, analysis of historical annual backflow test reports for industrial/commercial customers);
- investigative monitoring during initial implementation of a new or modified control measure (e.g. laboratory water quality testing of new pipe material to confirm the conformity to relevant standards); and
- adoption of technical data from published studies, including evidence of the effectiveness of

established industry best practices (e.g. evidence that installation of water meters with dual check valves prevents backflow from residential customers).

Validation related to significant risks will typically require assigning operational or critical limits, operational monitoring and corrective actions for violation of critical limits, as described in sections 6 and 8. It is essential to validate the critical limits to ensure that they are continuously effective in controlling the significant risks, and violation of critical limits will be significant to public health.

It should be noted that validation is uniquely different from operational monitoring (section 6) and verification monitoring (detailed in section 7). Operational monitoring determines whether control measures are working, whereas verification is required to confirm or reassure that the water quality delivered to consumers is safe and aesthetically acceptable.

The water supplier must be able to demonstrate, using historical water quality data, other monitoring data and operational procedures, that current control measures are effective in controlling the associated hazardous event. Typically, the control measures are validated when the WSP is developed for the first time and subsequently reviewed during annual and unscheduled audits of the WSP. Examples in Table 8 demonstrate how to validate control measures.

Table 8. Examples of validation of control measures

Hazardous event	Hazard	Control measure	Validation of control measures
Security breaches at water supply assets	M, P, C	• Security fencing, locked gates, access hatches, alarms, routine security patrols, closed-circuit television cameras	• Historical security inspection records indicate no security breaches in the past 2 years
Microbiological contamination of storage tanks due to faults in roofs, hatches, inlets, etc.	M, P, C	• Annual roof inspection programme	• Historical water quality data (*E. coli* results) and roof inspection reports indicate no breaches in roof integrity of all storage tanks in the past 12 months
Contamination during repair of ≤150 mm mains	M, P, C	• Dewatering of trench prior to commencing repairs • Prevention of contamination of pipe material during storage, transport and repairs • Flushing of water main – specify duration based on length of the main and minimum flow rate • Water quality testing (visual/turbidity) prior to turning on the water main	• Field audit reports for the past 12 months and checking of records indicate compliance with the repair procedure • Water quality data indicate that water complied with turbidity requirements and contained no *E. coli* after completion of repair • A study on water main repair practices that included microbiological testing after 50 main breaks indicated that the "burst repair procedures" effectively controlled microbiological contamination
Microbial contamination, growth of biofilm organisms	M	• Maintaining chlorine residual	• No *E. coli* detected in water samples, heterotrophic plate count numbers low • Published evidence of the effectiveness of chlorine in inactivating viral and bacterial pathogens and controlling biofilms (Olivieri et al., 1986; USEPA, 2002e,f)

C: chemical; M: microbiological; P: physical

4.3 Reassess and prioritize the risks

Most water suppliers adopt a simple semi-quantitative approach (section 3.3) for their risk assessment. It is common, but not essential, to assess the risks without considering the effect of current control measures. This "raw risk" assessment provides an insight into some hazardous events that can be a significant threat to public health if not adequately controlled. The next step in the risk assessment process is to reassess the risks considering the effectiveness of current control measures. This remaining risk is defined as the "residual risk".

The objective of reassessment of risks is to determine the effectiveness of existing control measures in preventing or removing significant risks. Residual risks should be prioritized from the highest to the lowest risk:

- High residual risk rating due to lack of or inadequate control measures – If a control measure is inadequate, improvements should be investigated, including enhanced management of the control, such as tighter critical limits, better alarm systems and quicker response times. Additional control measures should be included in an improvement plan (section 5).
- Medium and low residual risk rating – Operational monitoring of control measures for these events is essential to ensure that the level of risk remains low. The effectiveness of these control measures is generally monitored via the monitoring of related standard operating procedures.

Examples of the reassessment of risks are shown in Table 9, assuming that existing control measures are effective. The next section deals with responses if operational monitoring indicates that existing control measures are not effective.

4. DETERMINE AND VALIDATE CONTROL MEASURES, REASSESS AND PRIORITIZE THE RISKS

Table 9. Examples of reassessment of risks after application of control measures

Hazardous event	Hazard	Inherent or raw risk (risk ignoring the effect of controls)				Current control measure	Residual risk (if control measure effective)			
		Likelihood	Severity	Risk score	Risk rating		Likelihood	Severity	Risk score	Risk rating
Contamination of treated water storage reservoirs from birds and animals	M	3	5	15	High	Prevention of contamination and maintenance of chlorine residuals as described in Table 7	2	5	10	Medium
Security breaches at storage tanks	M, C	4	5	20	High	Storage area security (locked gates, alarms, remote cameras, routine inspections)	1	5	5	Low
Contamination from water main breaks	M	3	5	15	High	Prevention of contamination through applying appropriate procedures for repairing faults and returning the main to service (see Table 7)	1	5	5	Low
Entry of pathogens through backflow from illegal connections	M	2	5	10	High	No current control	2	5	10	High
Elevated DBPs	C	3	3	9	Medium	No current control	3	3	9	Medium
Taste and odour complaints due to sloughing of biofilms in water mains	P	2	3	6	Low	Routine water main flushing programme	1	3	3	Low
Increase in pH in concrete tanks	C	1	5	5	Low	Reservoir operating rules for seasonal variations to ensure maximum 48-hour filling cycle	1	5	5	Low

C: chemical; DBP: disinfection by-product; M: microbiological; P: physical

5. Develop, implement and maintain an improvement/upgrade plan

- Assemble WSP team **M1**
- Describe the water supply system **M2**
- Identify hazards & hazardous events, assess risks **M3**
- Determine and validate control measures, reassess and prioritize the risks **M4**
- **Develop, implement and maintain improvement plan M5**
  - Draw up improvement plan
  - Implement improvement plan
- Define monitoring of control measures **M6**
- Verify the effectiveness of the WSP **M7**
- Prepare management procedures **M8**
- Develop supporting programmes **M9**
- Plan and carry out periodic review of the WSP **M10**
- Revise the WSP following an incident **M11**

ENABLING ENVIRONMENT

If the previous step identifies that existing control measures are not effective or that necessary control measures are not present, then an improvement or upgrade plan needs to be identified. Improvement plans can include reviews of existing operating procedures for activities, implementation of new operating procedures or, in some cases, infrastructure changes to upgrade control measures, such as disinfection or maintenance of positive pressures. For example, in Table 9, entry of pathogens through illegal connections is identified as a high risk and formation of elevated concentrations of DBPs a medium risk due to lack of adequate control measures. In these cases, a cross-connection control programme needs to be included in an improvement plan. Elevated DBPs can be reduced either by reducing the concentrations of dissolved organic material in source waters or by reducing detention times of water in distribution systems. This could be achieved by eliminating sections of pipework that are not being used (dead ends), by increasing turnover through storage tanks during periods of low flow by dropping water levels or by taking storage tanks out of service during periods of low flow.

Similarly, if there are regular detections of *E. coli* in samples from distribution systems, despite control measures to minimize contamination of storage tanks, minimize the entry of contamination during installation of new mains or during repairs, etc., the performance of these control measures should be reviewed. One possibility could be to maintain chlorine residuals throughout the distribution network to provide additional protection against this high risk. This could require installation of secondary booster chlorination.

The first step in identifying improvements and upgrades is to consider options, including factors such as cost, practicality, timelines and likelihood of success. In the examples discussed above, some improvements could be achieved by improved operating procedures rather than more expensive treatment options. These should take priority. Where more than one improvement is required, priorities need to be identified, taking into account the level of unaddressed risk. Hence, reducing DBP concentrations would have a lower priority than preventing entry of pathogens.

Once options have been identified and priorities established, they need to be included in a comprehensive improvement plan. The plan should establish a schedule of short-, medium- and long-term activities. It is essential to also establish a mechanism for monitoring and reporting on implementation of the plan.

Other issues that need to be addressed include:
- responsibility for the improvement plan;
- financing;
- updating the WSP as improvements are introduced. This could require new or enhanced operational procedures and monitoring, updated incident protocols, training of operators, updated documentation and reporting; and
- verifying the effectiveness of improvements in reducing or eliminating risks.

6. Define monitoring of the control measures

Once the water in a drinking-water distribution system reaches the first consumer connection, it should be safe to drink without further treatment; therefore, it is important to maintain water quality and minimize the risk of contamination and deterioration of quality during transport.

The WHO *Guidelines for Drinking-water Quality* (WHO, 2011) advocate the use of WSPs as the most effective means of consistently ensuring the safety of drinking-water. A key aspect of the approach is an operational monitoring programme to indicate whether or not the control measures are operating within a target range. This section describes the types of parameters that are typically used in operational monitoring, the selection and justification of sampling locations and recommended frequencies of sampling.

6.1 Selection of appropriate operational monitoring parameters

Operational monitoring is a planned and routine set of activities used to provide timely indications of the performance of control measures and provide the opportunity for appropriate responses to non-compliance to maintain water quality. Such operational monitoring is usually based on simple observations and tests that can easily be measured and assessed, such as turbidity, chlorine residuals and infrastructure inspections that provide rapid feedback on how the system is working.

Operational monitoring has a different purpose from verification monitoring, which is limited in its ability to protect public health because consumer tap monitoring is typically (1) insufficient to provide early warning of contamination; (2) not indicative of what could have gone wrong between the treatment plant and the consumer's tap, so as to effectively guide remediation; and (3) too limited across space (too few sampling locations) and time (discrete small-volume samples are collected too infrequently) to provide information that applies to every potential user (National Research Council, 2006).

A principle of WSPs is that significant hazards are eliminated or minimized through collective application of control measures based on the multiple-barrier approach. Although distribution systems may differ in design, size and complexity, there are many common challenges, allowing generalizations to be made about their control. For example, maintenance of adequate chlorine disinfectant residuals to consumer connections and backflow prevention and cross-connection control are common control measures for drinking-water distribution systems (Table 10).

Under WSPs, for each control measure identified, an appropriate means of operational monitoring should be defined that will ensure that any deviation from required performance is detected in a timely manner (WHO, 2011). For some control measures, it is necessary to define "critical limits" outside of which water safety may be compromised and urgent action is required to rectify the problem. Examples of operational monitoring of control measures are provided in Table 10 for reference. The table includes parameters monitored, where they should be monitored, the frequency with which they should be monitored as well as the monitoring process.

6. DEFINE MONITORING OF THE CONTROL MEASURES

Table 10. Examples of operational monitoring of control measures used in distribution systems

Control measure	Target	What	Where	When	How	Who	Corrective action
Maintaining chlorine residual	• Minimum residual (e.g. 0.2 mg/L to all customers or percentage of system receiving at least 0.2 mg/L) • pH 7–7.5	• Chlorine residual • pH	• Customers' taps	• Multiple samples per week (weekly–monthly from individual taps)	• Field kit	Water quality officer/ sampling officer	• Adjust chlorine dose • Adjust pH
Booster chlorination (if used)	• Chlorine residual 2 mg/L at first monitoring point • pH 7–7.5 • Minimum residual to all customers (e.g. 0.2 mg/L)	• Chlorine residual • pH	• Monitoring point within 15–30 minutes of chlorinator based on flow	• Continuously (if possible) • Daily	• On-line • Field kit	Water quality officer/ sampling officer	• Adjust chlorine dose • Adjust pH
Leakage management	• Set water loss targets based on historical performance and performance of similar systems in region	• Percentage water loss	• Operations centre • Site inspections	• Annually	• Water audit • Monitor compliance with SOP for leakage management	Operations manager	• Review SOP
Maintaining positive pressure and flows	• Minimum pressure (e.g. 20 psi at any point in system; 50–75 psi at all residences) • Maximum 100 psi • Target flows based on historical performance • Avoid flow reversals and sudden surges of high flows • Minimize ingress of contamination and growth of biofilms • HPC maintained at set limits (no sudden increases) • Disinfectant residuals maintained (see above)	• Water pressure • Water flows • HPC • Disinfectant residuals	• Distribution system, including customers' taps	• Continuously (if possible) • Daily–weekly reading of meters and gauges (depending on location) • Multiple samples per week for HPC and residuals (weekly–monthly from individual taps)	• On-line • Reading of meters and gauges • Field kit for disinfectant • Laboratory analysis for HPC	Water quality officer/ sampling officer	• Identify cause of loss of pressure or flows • Institute remedial action to restore pressure and flows

Table 10 (continued)

Control measure	Target	What	Where	When	How	Who	Corrective action
Backflow prevention and cross-connection control	• No cross-connections or backflow into the distribution system	• Inspect new connections • Monitor testing of devices • Inspection programmes for existing devices • Where devices are installed in meters, monitor flows	• At premises	• Prior to connection • Annual review of device testing • Ongoing inspection programme • Quarterly meter reading	• Physical inspection of new connections • Review of device testing reports • Inspection of existing devices • Meter readings	Plumbing inspectors/ meter readers	• Replace or repair any faulty devices • Install devices where none fitted • Investigate cause of abnormal meter readings
Operating procedure for water main repair and installation of new mains	• Mains are repaired or installed following an SOP that minimizes contamination • No E. coli introduced into the system, no increase in turbidity, disinfectant residuals maintained after completion, maximum residuals not exceeded	• Water quality following repair or installation of new main (e.g. E. coli, turbidity, disinfectant) • Compliance with SOP	• On-site • Operations centre	• At completion of repair/installation	• Site inspection • Check documentation to ensure compliance with SOP, including results from water quality monitoring of residual, pH	Operations manager/ works supervisor	• If E. coli detected, undertake disinfection, flushing and repeat sampling • Take further samples to determine extent of contamination • If high turbidity detected, undertake water main flushing until normal turbidity levels restored

6. DEFINE MONITORING OF THE CONTROL MEASURES

Table 10 (continued)

Control measure	Target	What	Where	When	How	Who	Corrective action
Quality control for chemicals and materials	• All materials are suitable for contact with drinking-water and do not lead to contamination of drinking-water • Chemicals comply with quality requirements	• All materials certified for contact with drinking-water • All batches of chemicals accompanied by analytical results meeting quality requirements	• Operations centre • At receiving sites (depots/stores)	• Prior to purchasing materials and chemicals and on receipt • Annual review of records for compliance	• Check documentation, including certification of materials and chemical analytical results for each batch of chemicals received	Water quality officer/procurement manager	• Do not accept unsuitable materials • If suspected that they have been installed, undertake immediate water quality testing to determine impact • Discard non-compliant chemicals and replace materials if possible
Maintenance of storage tanks, fittings and mains	• Integrity of system maintained to prevent ingress of contamination • Minimize accumulation of sediments and growth of biofilms • HPC maintained at set limits (no sudden increases) • Disinfectant residuals maintained (see above)	• Integrity of infrastructure, performance of equipment, cleanliness of system • HPC • Disinfectant residuals	• On-site • Sampling locations in distribution system	• As specified in documented programme for maintenance and inspection activities (e.g. storage tank integrity, cleaning of storages, water main flushing) • Multiple samples per week for HPC and residuals (weekly–monthly from individual taps)	• Physical inspection of infrastructure, including storages • Cleaning of storages • Water main flushing • Checking operation of valves and fittings	Operations and maintenance personnel	• Repair faults in storage tanks • If disinfectant residuals are low, investigate mechanisms to reduce detention times • If excess sediment/biofilms detected, undertake cleaning and disinfection
Maintaining asset security	• No unauthorized access to infrastructure • No interference with water quality	• Security barriers maintained	• Site inspections • Operations centre (alarms, CCTV, if installed)	• Ongoing as specified in documented programme • Continuous monitoring of alarms and CCTV (if installed)	• Physical inspection of infrastructure (fences, locks, storage hatches, etc.)	Operations and security personnel	• Repair security barriers • Test water and, if possible, take affected infrastructure out of supply • Review security procedures and monitoring

CCTV: closed-circuit television; HPC: heterotrophic plate count; psi: pounds per square inch (1 psi = 6.9 kPa); SOP: standard operating procedure

6.1.1 Types of parameters

Desirable attributes for operational monitoring parameters include speed and ease of measurement, low cost and ability to be monitored continuously or regularly across the water distribution network. Monitoring of certain parameters may also be required to meet regulatory requirements; typically, these are among the more useful and informative parameters (e.g. chlorine residual, turbidity, pressure changes) (National Research Council, 2006).

Parameters should be selected with an understanding of the possible mechanisms that could be responsible for changes in water quality (e.g. turbidity changes associated with flow reversals, which in turn could trigger sloughing of biofilms).

Operational monitoring should be performed at a frequency that enables timely intervention before control of water quality is lost and unsafe water is delivered to consumers. Observational monitoring, such as inspection of water storages, will be undertaken less frequently than testing of water quality parameters. Parameters that can be monitored on-line provide distribution system managers with real-time opportunities to implement operational controls. Monitoring devices can be set to trigger alarms at alert levels that are within critical compliance limits but allow timely interventions to bring the system back into the desired operating range. Most physical parameters (e.g. flow, storage tank levels, turbidity) tend to be relatively inexpensive to monitor and reliable for continuous monitoring. Real-time monitoring of inorganic and organic chemicals and biological organisms is more restricted. Biological monitoring devices are limited to alarm-type systems that detect behavioural changes in a microcosm population of fish or invertebrates or fluorescence changes in algae in response to changes in water quality.

Sentinel parameters for monitoring distribution system integrity are listed in Table 11.

Table 11. Sentinel parameters for distribution system integrity

Parameter[a]	Physical	Hydraulic	Water quality	Biological
Routine (primary)				
Pressure	✓	✓		
Turbidity	✓	✓ (flow reversals)	✓	
Disinfectant residual		✓ (water age)	✓	
Main breaks	✓			
Water loss	✓			
Colour	✓ (corrosion)		✓	
Coliforms/E. coli	✓ (sanitary, main break)		✓ (biofilms)	
Flow velocity and direction		✓ (pipes, tanks)		
pH, temperature			✓	
Chemical parameters	✓	✓	✓	✓ (if toxic)[b]
Secondary				
Total organic carbon			✓	
Ultraviolet absorption			✓	
Trace organic compounds, including DBPs			✓	✓ (if toxic)
Taste and odour	✓ (permeation)	✓ (water age)	✓ (biofilms)	✓ (if toxic)
Metals	✓ (corrosion)		✓	✓ (if toxic)
Nitrite/nitrate			✓ (nitrification)	
Heterotrophic plate count bacteria			✓ (biofilms)	
Tank level/volume		✓		

DBPs: disinfection by-products

[a] Bold parameters are those for which on-line real-time sensors are available.
[b] Excludes chlorine, as testing can be done only after dechlorination to protect the biota being monitored.

Source: After National Research Council (2006)

6. DEFINE MONITORING OF THE CONTROL MEASURES

6.1.2 On-line operational monitoring

6.1.2.1 Disinfectant residual

On-line real-time disinfectant residual monitors can measure free chlorine, chloramines or oxidation–reduction potential. The technologies employ polarographic, voltammetric or colorimetric methods that can influence the device sensitivity, calibration and interference from other water quality parameters (National Research Council, 2006). The free chlorine concentrations at the customer's tap should preferably be in the range of 0.4–0.6 mg/L for aesthetic reasons and always below the health-based guideline value of 5 mg/L. Chlorine doses should be managed to achieve effective disinfection while minimizing the formation of DBPs (WHO, 2011).

6.1.2.2 Flow

Flows can be influenced by pumping regimes, storage tank operations and manipulations of hydrants or blow-off valves. Monitoring of flows using in-line meters is typically conducted at sub-district boundaries to provide comparisons with customer meter data and so allow measurement of leakage rates. A distribution system hydraulic model can make use of the flow data to generate detailed descriptions of distribution system water velocities and flow reversals.

6.1.2.3 pH

A wide variety of on-line glass electrode pH meters are available. Measurements are reliable, but regular calibrations are required to avoid drift. An optimal pH range for drinking-water distribution systems is normally 6.5–8.5 (WHO, 2011). Drinking-water pH can increase through distribution systems due to leaching of lime from concrete storage reservoirs and cement-lined pipes (see section 3.2.3), with the amount of increase proportional to the detention time of the water within the distribution network. Chlorine efficacy is optimal at about pH 7.0, decreases significantly with increasing pH and will be ineffective at pH 10.

6.1.2.4 Pressure

Operational monitoring of transient pressure changes using high-speed electronic pressure data loggers is recommended by the United States National Research Council (2006). High-speed devices (sampling up to 20 times per second) are necessary, because distribution system pressure transients may last for only a few seconds and may not be observed by conventional pressure monitoring. Pressure detection units are programmable and can be set to trigger alarms at specific thresholds.

6.1.2.5 Temperature

Temperature thermistors typically work over a relatively small temperature range and can be very accurate within that range. The measurements are very reliable and typically do not require routine calibrations (National Research Council, 2006). Temperatures in excess of 20°C may be a concern for free chlorinated systems to maintain a residual and because of potential growth of opportunistic pathogens (section 3.1.1).

6.1.2.6 Turbidity

Suspended sediments and corrosion products such as iron and manganese can cause elevated turbidities. There are many different models of on-line real-time turbidity meters available. However, in the finished water distribution system, turbidity probes need to be sensitive at low ranges (i.e. <1 nephelometric turbidity units); therefore, the more sensitive low-range devices are preferred.

On-line particle counters set to specific ranges, such as for particles 2–15μm in size, may provide more useful indication of pathogen breakthrough of filtration barriers, but probably provide no improvement over turbidity to detect pathogen ingress into distribution systems, given the high number of similarly sized soil and corrosion products in distribution systems.

6.1.2.7 Chemical parameters

Ion-selective electrodes can be used to monitor analytes such as chloride, nitrate and ammonium ions, among others. However, these devices are not always ion specific, and ionic interference may influence monitoring results. Spectrophotometers that record the percentage of ultraviolet light absorbed by water report a rate of ultraviolet transmittance. As double bonds and ring structures strongly absorb light at 254nm, ultraviolet transmittance can be used to determine the amount of organic matter present that can contribute to colour in water. Commercial total organic carbon on-line monitors are available, although typically such devices have higher maintenance and operating requirements compared with other devices.

Single- or multi-parameter sensors are widely available that can communicate directly with supervisory control and data acquisition (SCADA) systems. Integration of monitoring into SCADA systems provides an effective means of rapidly processing and responding to the large quantities of data generated by on-line monitoring devices (large drinking-water distribution networks in particular will involve the generation of very large quantities of on-line data).

6.1.2.8 Biological monitoring devices

In recent years, more commercial biological monitoring devices have become available. Other names in use include "biomonitors" and "toximeters". The devices typically consist of a microcosm population of organisms that respond to sudden changes in water chemistry through changes in patterns of activity that can be detected by a sensor alarm. Such devices essentially work as a broad-spectrum chemical alarm. As no information is provided by the devices on what triggered the change in activity of the monitored organisms, biological monitoring devices are generally used only to monitor the quality of treated water entering treated water storage tanks before its release to the distribution system. Alarms can trigger shutoffs or switching of supply tanks.

6.2 Reviewing operational monitoring data

Review of operational monitoring data should occur at a range of frequencies, depending on the data collected and their purpose. On-line monitoring data are usually alarmed and, for larger systems, connected to SCADA systems. Programmable logic controller devices can be programmed to trigger alarms to operators when alert levels or critical limits are hit for individual parameters or when combinations of events or trends occur. These data should be reviewed over longer terms to identify patterns of data that can be considered as precursors to poor water quality. The frequency of review depends on the parameters and the experience of the operator and supervisors, but periodic weekly and monthly reviews of operational monitoring data trends should be considered.

Observational data, such as inspection of water storages and records from standard operating procedures, such as those applied when installing or repairing mains or in conducting quality assurance of chemicals and materials, will be generated with a lesser frequency. Immediate responses may be required when significant faults, such as gaps in storage roofs, are detected or contamination is detected following repairs to damaged mains. In these cases, reviews of reports should be undertaken immediately. In addition, longer-term reviews should be undertaken on at least an annual basis.

Over the longer term (e.g. annually), operational data should be reviewed as part of the water supply system WSP periodic review (see Bartram et al., 2009). Review of operational data trends can assist in identifying problem points in system operations and in prioritizing capital improvement needs. The annual review of operational data also plays a role in the ongoing verification of compliance and validation of existing controls.

7. Verify the effectiveness of the WSP

- Assemble WSP team
  M1
- Describe the water supply system
  M2
- Identify hazards & hazardous events, assess risks
  M3
- Determine and validate control measures, reassess and prioritize the risks
  M4
- Develop, implement and maintain improvement plan
  M5
- Define monitoring of control measures
  M6
- **Verify the effectiveness of the WSP**
  M7
- Prepare management procedures
  M8
- Develop supporting programmes
  M9
- Plan and carry out periodic review of the WSP
  M10
- Revise the WSP following an incident
  M11

ENABLING ENVIRONMENT

- Set up verification monitoring programmes for microbial and chemical parameters (include on-line parameters)
- Monitor customer satisfaction
- Conduct internal and external audits

Verification is the final check of water safety. It provides an objective confirmation of the overall safety of the system and that the WSP is working effectively, as well as identifying issues for improvement. Verification is a set of review and audit processes – in other words, although corrective actions will be required to address faults or shortcomings, verification is not used as a short-term control measure. Faults identified by verification will nearly always be identified after water has been delivered to consumers.

Verification involves three separate activities:
- monitoring the quality of drinking-water supplied to consumers;
- monitoring consumer satisfaction; and
- internal and external auditing of operation of the WSP.

In combination, these activities will determine whether the WSP is being implemented as intended and whether it is functioning effectively in supplying water to consumers that complies with water quality requirements.

7.1 Verification monitoring

Verification monitoring involves testing the water supplied to consumers to determine compliance with water quality targets identified in regional or national guidelines or standards and specified in the WSP and any customer contracts with respect to finished product requirements. Verification monitoring fulfils a different purpose from operational monitoring (section 6). Operational monitoring determines whether individual control measures are working, whereas verification monitoring determines whether the collection of control measures combined within a WSP have been effective. Verification monitoring involves different parameters that are generally monitored at lower frequencies. The parameters typically include *E. coli* as an indicator of microbial quality and health-related chemicals (see sections 7.1.1 and 7.1.2).

Verification monitoring programmes should identify what parameters will be tested and where, when and by whom samples will be collected. Verification monitoring can be performed by water suppliers, regulatory agencies (as part of surveillance) or both. In some cases, water suppliers may choose to have samples taken and tested by independent agencies for external quality assurance. If more than one agency is involved in verification monitoring, it is important that the various programmes are coordinated to ensure that they are complementary rather than being unnecessarily duplicative. Verification monitoring should be consistent with regulatory requirements in terms of both the range of parameters and the frequency of testing. The range of parameters will normally be based on consideration of the nature of the source water (e.g. groundwater or surface water), catchment activities (e.g. presence of industry or agriculture), water treatment processes, type of disinfectants and construction of distribution systems. Frequency will be based on expected variability in concentrations of individual parameters, as well as the size and complexity of the distribution system and the population served by it.

Other issues that need to be considered and identified include documented sampling methods, training of samplers and availability of appropriate laboratories. Where possible, laboratories accredited by national certification schemes should be chosen.

It is essential that result review processes, responses and communication protocols should be established. These should include responses to results that do not comply with guideline values, standards and regulatory requirements. Responses can range from immediate resampling to boil water and avoid consumption advisories (section 8.2). Unless there are unusual circumstances, it is unlikely that a single non-compliant result (e.g. detection of *E. coli* in a single sample) will lead to a public notification or warning being issued. The first step will normally be an investigation of potential causes and prompt

7. VERIFY THE EFFECTIVENESS OF THE WSP

collection of additional samples. The investigation should consider the operation of control measures both within the distribution system and upstream. Communication protocols need to include internal and external reporting processes. Potential responses from external agencies to non-compliance should be discussed and, as far as possible, agreed.

7.1.1 Microbial parameters

This section focuses on the verification monitoring of microbial indicator organisms to assess the integrity of the distribution system.

The WHO *Guidelines for Drinking-water Quality* (WHO, 2011) recognize *E. coli* as the indicator of choice for faecal contamination, although thermotolerant coliforms (*E. coli*, *Citrobacter*, *Klebsiella* and *Enterobacter*) can be used as an alternative. Thermotolerant coliforms are less specific indicators, as strains can grow in the environment and may not be of faecal origin. Although *E. coli* is useful, it has limitations. Enteric viruses and protozoa are more resistant to disinfection; consequently, the absence of culturable *E. coli* will not necessarily indicate freedom from these organisms. Under certain circumstances, the inclusion of more resistant indicators, such as bacteriophages and/or bacterial spores, should be considered (WHO, 2011). *Escherichia coli* also provides no indication as to the presence of environmental pathogens (e.g. *Legionella* and *Naegleria fowleri*).

Many jurisdictions monitor total coliforms within the distribution system and at customers' taps. Total coliforms are not a specific indicator group for contamination, as coliforms can grow naturally in water and soils. However, they can be used to assess the cleanliness of distribution systems. Coliforms can arise from biofilm linings in pipes and fixtures or from contact with soil due to breaks or repair works. Testing for heterotrophic plate count bacteria is sometimes used for similar purposes and may provide a more sensitive measure, given their high numbers in soils.

The presence of and/or trends in total coliform numbers and heterotrophic plate count bacteria are used in operations as an indicator of system performance and may forewarn of potential system problems, including a loss of disinfection efficacy, intrusion of contaminants into drinking-water or the growth of biofilms that could support the presence of pathogens such as *Legionella* and *N. fowleri*. The recurring detection of any coliforms should trigger a corrective action, such as increasing the chlorine dose at the water treatment plant, checking the operation of service reservoirs or pipe flushing and rechlorination of the affected area.

7.1.2 Chemical parameters
7.1.2.1 Health-related chemicals
Health-related chemicals include:
- naturally occurring inorganic chemicals, such as arsenic, boron, chromium, fluoride, molybdenum, selenium and uranium;
- industrial chemicals, such as benzene, cadmium, cyanide, mercury, styrene, toluene and xylene;
- contaminants from pipes and fittings, such as antimony, benzo(a)pyrene, cadmium, copper, lead, nickel and vinyl chloride;
- agricultural chemicals, such as nitrates and pesticides;
- water treatment chemicals, such as aluminium, chlorine and chloramines; and
- DBPS.

The selection of chemicals in verification monitoring will be informed by knowledge of the water system and hazard identification, as described in section 3. The potential presence of health-related chemicals at concentrations exceeding water quality targets will depend on source water quality, pipe and fitting materials used in the distribution system, agricultural activities and use of pesticides in water catchments, and water treatment chemicals used, including the type of disinfectants.

For example, THMs and HAAs are considered good indicators for the majority of chlorination by-products. However, other DBPs should be included in verification monitoring programmes when other types of disinfectant are used (e.g. NDMA when chloramination is used).

7.1.2.2 Metals that influence acceptability

It is common practice to test for metals, such as iron, manganese and copper, for aesthetic reasons. Iron is typically present in the water supply from the corrosion of iron or steel pipes or other components of the plumbing system, whereas manganese is common in dissolved mineral form in surface waters and groundwaters and can give rise to undesirable tastes and staining of clothes during washing. Copper arising from copper plumbing and fixtures may occur at high concentrations at customers' taps, particularly if the water has remained stagnant in the plumbing system for long periods.

Increasing concentrations of metals can be used as a trigger for water main cleaning to avoid customer complaints.

7.1.3 Example distribution system verification monitoring programme

Verification monitoring includes some parameters that may also be tested on-line elsewhere in the distribution system as part of operational monitoring (e.g. chlorine, pH). Testing frequencies for verification will vary depending on the parameter, but testing is typically less frequent than testing of parameters used in operational monitoring. An example suite of testing parameters is shown in Table 12.

Table 12. Example of parameters for verification monitoring at customers' taps

Chemical	Physicochemical	Microbiological
Chlorine	pH	Total coliforms
Arsenic	Temperature	*E. coli* (or thermotolerant coliforms)
Fluoride	Turbidity	Heterotrophic plate count bacteria
Hardness	Conductivity	*Legionella pneumophila* (infrequent, if deemed an at-risk system)
HAAs	Alkalinity	*Naegleria fowleri* (infrequent, if deemed an at-risk system)
Nitrate	Total dissolved solids	
Selenium	Colour	
THMs		
Metals		

HAAs: halogenated acetic acids; THMs: trihalomethanes

7.1.4 Choosing sampling locations

7.1.4.1 Designating sampling zones

Larger distribution networks generally require more samples to characterize water quality due to greater differences in network attributes such as flow rates, water retention times, pipe material and pipe age. Larger networks may also receive water from different service reservoirs, and there may be distinct geographical discontinuities, such as two suburbs or towns separated by a major road or river. A common approach is to split larger networks into zones or subdistricts with the view to conducting a verification monitoring regime within each zone that effectively characterizes water quality in that zone.

7.1.4.2 Selection of sampling sites

A common practice among water utilities is to rotate among designated sampling sites across the distribution system. Here, the aim is to characterize water quality within the zone effectively and enable comparisons of water quality over time for particular sections of the system. Rotation of sampling sites avoids the problem of sampling from the same site each time, which could give a misleading characterization of water quality. It is important that the sampling frequencies and locations are selected to provide the greatest confidence that all parts of the system are operating within the target ranges and, in the case of certain microbial parameters, free from contamination.

The location of sample points across each zone should reflect the number of people served (this is particularly important for microbiological samples). Different parts of the zones may include branch pipelines or loops, different pressure zones or areas receiving water from different sources or different treatment plants.

A day-to-day water verification monitoring programme involves sampling and testing from many locations throughout the distribution network. This commonly includes use of purpose-built sampling fittings located at the boundary with customers' properties (often referred to as "customers' taps"). Not all designated sampling sites need to be sampled on each sampling occasion; rather, a rolling programme where fixed sites plus a selection of randomly chosen sites (e.g. customers' taps) are sampled intermittently is recommended. Augmenting the fixed-site programme with randomly selected sites avoids the risk that changes may arise over time in parts of the system that fall between the fixed sites.

7.1.5 Sampling frequency

Samples should be collected at regular intervals throughout the annual sampling calendar. The more frequent the sample collection, the greater the confidence that sample results will effectively characterize the true water quality within the system. The WHO *Guidelines for Drinking-water Quality* (WHO, 2011) provide guidance on the minimum number of samples for faecal indicator testing in distribution systems (Table 13). Many jurisdictions also provide their own guidance.

Table 13. Recommended minimum sample numbers per year for *E. coli* testing in piped distribution systems

Population	Total number of samples per year
<5 000	12
5 000–100 000	12 per 5 000 population
>100 000–500 000	12 per 10 000 population plus an additional 120 samples
>500 000	12 per 50 000 population plus an additional 600 samples

[a] Parameters such as chlorine, turbidity and pH should be tested more frequently as part of operational and verification monitoring.

Source: WHO (2011)

The chances of detecting contamination in systems reporting predominantly negative results for faecal indicator bacteria can be increased by using more frequent presence/absence testing (WHO, 2011). Presence/absence testing can be simpler, faster and less expensive than quantitative methods; however, its use is appropriate only in systems where the majority of tests for indicator organisms are negative.

The more frequently the water is examined for faecal indicator organisms, the more likely it is that contamination will be detected. Frequent examination by a simple method is more valuable than less frequent examination by a complex test or series of tests (WHO, 2011).

7.2 Customer satisfaction

Verification includes monitoring consumer satisfaction. This type of verification is often overlooked or undervalued; however, it can be very powerful in detecting faults and measuring improvement. This is particularly true for aesthetic water quality problems in distribution systems.

One method for performing this activity is to establish consumer communication and response procedures and to monitor and document complaints and feedback. These should be analysed and reported to senior management. Patterns of complaints should always be investigated. Although consumers are subjective and untrained, they can provide reliable reports of water quality problems that enable more rapid follow-up investigation and maintenance by the water utility. There are many examples of situations where consumer complaints and feedback have identified contamination incidents. Outbreaks in Naas (Ireland), Fife (Scotland) and Brushy Creek (Texas, USA) were all detected following customer complaints (Hrudey & Hrudey, 2004). Discoloured water, increased turbidity and off-odours can provide evidence of ingress of contamination through backflows from cross-connections, water main breaks and other faults. Recording patterns and frequencies of consumer complaints using a geographic information system–linked database is commonly undertaken to assist in the operational tracking of the water quality issue and identifying the boundary of the affected area.

7.3 Internal and external auditing

The third form of verification is auditing of compliance with the WSP. Audits will generally involve interviewing managers and operational staff. The objective is to assess that the plan is being implemented as intended and is effective.

Audits will normally include:
- checking that the description of the distribution system is accurate;
- checking that significant hazards and hazardous events have been identified and that the risk assessment was logical and thorough;
- reviewing measures and activities designed to monitor and manage potential impacts of connected buildings and facilities;
- assessing that appropriate operational monitoring was undertaken, that results were kept within set limits and that appropriate action was taken to respond to non-compliance;
- reviewing all operational procedures associated with the maintenance and repair of distribution systems (e.g. repair of water main bursts) to ensure that they are designed and implemented to reduce risk of contamination of distribution systems (see section 8);
- ensuring that verification monitoring programmes are in place and that results demonstrate that the WSP was effective;
- reviewing responses to incidents and emergencies and application of corrective actions;
- assessing implementation of improvement programmes and adoption of training plans;
- assessing the performance of subcontractors and management;
- ensuring that reporting requirements have been met;
- checking that all activities and results have been documented; and
- ensuring that regulatory requirements have been met.

An audit report should be prepared at the completion of each audit, describing findings, including recommended improvements or remedial measures, together with timelines. Findings should be discussed with the drinking-water provider, and a copy of the report should be provided. Audits may be internal or external processes, and they provide important input to the periodic review of the WSP (see section 10).

7. VERIFY THE EFFECTIVENESS OF THE WSP

7.3.1 Internal audits

Internal audits should be based on a peer principle – that is, auditor and auditee are considered as peers during the audit, irrespective of their formal positions, to facilitate a positive and confident atmosphere. Internal auditing is about continuous improvement, not about blame and fault. The audit should be performed according to an agreed programme to ensure that all parts of the WSP are audited regularly and prior to the regular WSP review.

7.3.2 External audits

External audits may form part of independent surveillance (see section 12) and be carried out by regulatory agencies, certification companies or independent experts, depending on the situation. The external audits are third-party assessments of the WSP, to provide independent documentation for compliance with regulatory requirements, consistency with standards or coherence with good practice. Additionally, the external audit provides credibility in relation to public conception of water safety.

8. Prepare management procedures

- Assemble WSP team **M1**
- Describe the water supply system **M2**
- Identify hazards & hazardous events, assess risks **M3**
- Determine and validate control measures, reassess and prioritize the risks **M4**
- Develop, implement and maintain improvement plan **M5**
- Define monitoring of control measures **M6**
- Verify the effectiveness of the WSP **M7**
- **Prepare management procedures M8**
- Develop supporting programmes **M9**
- Plan and carry out periodic review of the WSP **M10**
- Revise the WSP following an incident **M11**

ENABLING ENVIRONMENT

- Prepare and regularly review, test and revise standard operating procedures
- Prepare and regularly review, test and revise incident protocols

Clear management processes in the form of standard operating procedures (SOPs) document actions that are taken when the system is running under normal conditions. In addition, while the aim is that procedures will always be effective, incidents and events will occur, even in the best managed systems. Incidents and events represent deviations from defined operating limits and practices. Operational responses need to be developed to deal with such incidents. In effect, each SOP will describe normal operational practices, whereas incident protocols will describe actions and responses to be applied when the practices fail.

Reviews of SOPs, including critical assessments of risks and vulnerabilities, can predict many types of potential deviations, such as water main breaks or cross-connections. It is essential that plans and procedures should be developed for each of these events and included in WSPs to guide responses. Generic incident protocols should also be developed to deal with unforeseen events or incidents. These will include clear identification of roles and responsibilities, reporting requirements and communication procedures. Incident protocols should include alternative sources of water (including bottled water and carted water) that could be required in the event of prolonged interruptions to supply following substantial contamination incidents.

It is essential that SOPs and incident protocols are regularly reviewed, tested and revised, as necessary. This is most effective when operational staff are involved. Mechanisms should be established to ensure that managers and operations staff are provided with the most recent versions of procedures and protocols. These personnel need to be appropriately trained to implement the procedures.

8.1 Standard operating procedures

SOPs for distribution systems typically include procedures and processes for:
- maintaining flows and positive pressure and minimizing surges;
- operating intermittent supplies;
- maintaining disinfection throughout the distribution system;
- mixing water supplies from different sources;
- inspection and maintenance of storage tanks, service reservoirs, valves and other fittings;
- water leakage management;
- preventing corrosion;
- selection of pipe materials and chemicals connecting new customers, including selection and installation of backflow prevention devices;
- ongoing evaluation of backflow prevention devices and community education on backflows and cross-connections;
- repairing water main breaks;
- construction and commissioning of new mains;
- dewatering and recharging distribution mains;
- controlling permeation;
- collection and testing of water samples (what, where, when, how and who);
- calibrating equipment and SCADA systems; and
- dealing with customer enquiries.

Other SOPs could be required, depending on specific characteristics of individual disinfection systems.

Guidance on many of these procedures is provided in *Safe Piped Water: Managing Microbial Water Quality in Piped Distribution Systems* (Ainsworth, 2004).

8. PREPARE MANAGEMENT PROCEDURES

8.1.1 Positive pressure and adequate flows

The aim is to supply water at adequate pressure and flow. Loss of pressure can allow ingress of contamination (LeChevallier et al., 2003; Hunter et al., 2005) and can exacerbate impacts from cross-connections with inadequate backflow protection devices. Pressure at any point in the distribution system needs to be maintained within a range to avoid pipe bursts due to high pressure while maintaining minimum flow rates at all expected demands. In some jurisdictions, water utilities are required to provide water at minimum pressures on property boundaries. For example, a minimum of 20 pounds per square inch (psi)[1] at all points in the distribution system, a range of 50–75 psi at all residences and a maximum of 100 psi could be specified (Kirmeyer et al., 2001).

Asset system design should ensure that all components (e.g. controls for pump stations and pressure-regulating valves) that can influence water pressure and flows are managed in a coordinated plan to ensure that adequate pressures and flows are maintained and pressure surges are avoided. Diurnal and seasonal variations in drinking-water demands as well as impacts of sudden changes in water flow associated with activities such as firefighting need to be considered.

SOPs should include mechanisms for operational monitoring of pressures within distribution systems and protocols for investigating loss of positive pressure, especially during system changes and repair work.

SOPs should also include measures to maintain adequate flows through systems, particularly at ends of mains and during periods of low demand. Increased water age can lead to increased concentrations of DBPs, decreased disinfectant concentrations, nitrification in chloraminated systems, unacceptable tastes and odours due to biological growth, and low dissolved oxygen and increased pH in cement or cement-lined pipes.

Asset control systems and SOPs should avoid measures that lead to rapid increases in flows and rapid flow reversals, as both can dislodge accumulated sediments and biofilms, leading to increased turbidity and colour in drinking-water supplied to consumers.

8.1.2 Intermittent flows

Intermittent supplies present greater risks for entry of contaminants due to low pressures, backflows and physical faults. Although such systems are not ideal, they are a reality for a large proportion of the world's population. SOPs should include regular inspections for sources of contamination in the immediate vicinity of pipes and for signs of leakage. In addition, SOPs should apply to management of pumps and valves to ensure that minimum positive pressures are maintained when water is being supplied.

8.1.3 Maintaining disinfectant residuals

Maintenance of disinfectant residuals within distribution systems is used as a barrier following intrusion of bacterial and viral pathogens into distribution systems and as a mechanism to reduce the formation of biofilms and the growth and persistence of free-living pathogens such as *Legionella* and *Naegleria fowleri*. Although residual disinfectant will provide some protection, it should not be relied upon to deal with large events, as these will often quickly consume available disinfectant. Procedures for maintaining residuals should be used in conjunction with SOPs for minimizing external contamination through cross-connection control, pressure management, infrastructure maintenance, etc.

---

[1] 1 psi = 6.9 kPa.

Provision of disinfectant residuals is often a balance between avoiding excessive concentrations in water delivered to customers at the start of distribution systems and maintaining detectable concentrations at or near the ends of distribution systems, particularly when chlorine is used. Maintaining disinfectant residuals throughout distribution systems is not always possible; in these cases, targets should be set defining the extent of the system receiving measurable disinfectant. In some cases, secondary booster chlorination stations located within distribution systems are used to extend the delivery of disinfectant residuals. In other cases, particularly where distribution systems include long pipelines and extended resident times, chloramination has been used to improve persistence.

Whatever disinfection processes are used, SOPs should be developed to ensure that disinfection targets are achieved. SOPs should define target criteria, such as minimum and maximum residuals at the head of distribution systems or immediately after booster stations and target residuals in the distribution system. SOPs should deal with the operation of primary and secondary disinfection stations, operational monitoring requirements and responses to sudden drops in disinfectant residuals that could indicate contamination events. Where chloramination is used, strategies to minimize nitrification need to be developed, as well as corrective actions if nitrification occurs.

SOPs also need to deal with control of DBPs. While maintaining effective disinfection has the highest priority and should not be compromised, it is also important to minimize the formation of DBPs.

8.1.4 Mixing water sources

Where water distribution systems can receive water from a number of sources, SOPs should be developed to deal with mixing of these sources and, in particular, with changes in sources. Changes can lead to a range of issues, which can primarily affect the aesthetic and chemical quality of water, including dislodging of biofilms and deposits, loss of disinfectant residuals and change in taste. SOPs should deal with potential issues based on consideration of differences in chemical quality, pH, disinfection regimes and frequencies of mixing events. Where aesthetic changes or impacts on specialist users are likely, the SOP should deal with communication with consumers.

8.1.5 Inspection and maintenance of storage tanks/service reservoirs, valves and other fittings

Regularly programmed inspection and maintenance of distribution systems are an essential requirement for ensuring sustained performance. Structural and functional deficiencies can lead to loss of water pressure and ingress of contaminants, whereas accumulated sediments can lead to aesthetic issues. SOPs should describe the planned programme of maintenance activities, maintenance requirements, responsibilities for undertaking maintenance and communication requirements. Communication is particularly important where activities can lead to pressure loss, interruption to supply or the possibility of discoloured water (e.g. due to resuspension of sediments caused by changes in flow patterns).

All accessible infrastructure, including storage tanks, service reservoirs, pumps, valve chambers and above-ground pipes, should be inspected and maintained on a regular basis. Inspections should include examination of physical integrity and functionality (Table 14).

8. PREPARE MANAGEMENT PROCEDURES

Table 14. Examples of inspection criteria for service tanks, pumps, valves and hydrants

Inspection criteria	
Service tanks	**Pumps, valves and hydrants**
• Damage to roofs, including gaps in covers and hatches • Gaps between the roof structure and the tank wall • Gaps at entry points of pipework or cables through the tank roof or wall • Inadequate protection of overflows and vents to prevent small-animal access • Signs of ponding on roofs, indicating poor drainage • Cracks in tank walls (concrete tanks) or signs of corrosion (metal tanks) • Evidence of animal entry and nesting/roosting of birds on internal structures • Corrosion of internal pipework and structural components • If membranes are used, check for damage, splits in joints and cracks at edges • If the tank is lined or coated, check for damage to the lines or gaps in coatings • Accumulation of sediment • Localized growth of grasses or wet soil in the vicinity of tank walls, indicating leakage • Leaks from valves and external pipework • Security measures, including fences and padlocks, are in place and undamaged	• Leaks from seals • Accessibility (i.e. that they have not been covered by earthworks) • Cleanliness and dryness of housings (i.e. no evidence of leakage) • Functionality, including sealing capacity when in closed positions, number of turns required to open or close valves, operation of hydrants[a] • Security measures, including enclosures and padlocks, are in place and undamaged

[a] Care needs to be taken that operating equipment does not have unintended consequences on pressures and flows. For example, valves that have not had settings changed for a long time may break when operated; opening of closed valves may result in sediments lodging in the valve seat, preventing reclosure.

In addition to inspections, maintenance is required to ensure that infrastructure is kept clean of sediment and that devices such as pumps, valves and hydrants continue to operate effectively.

Cleaning of tanks should be undertaken at regular intervals (e.g. every 1–5 years) based on accumulation of sediments. Preplanning of cleaning is required to deal with impacts on water flows and pressures. Taking tanks off-line for cleaning can change the direction of water flows in local distribution networks, leading to dislodging of sediments from pipes and affecting the aesthetic quality of water delivered to consumers. Tank cleaning will normally include removal of sediments, pressure jetting of internal surfaces and disinfection with moderate doses of chlorine (10–20 mg/L). SOPs need to deal with disposal of sediments removed from tanks. If chemicals such as dilute acid solutions are used in cleaning, they need to be suitable for use in drinking-water systems. Prior to returning tanks to service, it is important to test water in the tank for pH, free chlorine levels and, wherever possible, microbiological quality.

Some water utilities also undertake regular flushing of mains. This may be targeted towards parts of distribution systems that are prone to collection of sediments and have a history of dirty water complaints or off-odours due to long detention times and low flows (e.g. distal locations, ends of branch mains). A range of methods can be used, including unidirectional flushing, air scouring and swabbing. SOPs need to be developed for each method. Preplanning is required for water main flushing programmes to deal with potential impacts on water pressure and flows. SOPs need to deal with disposal of sediments and dirty water generated by flushing.

SOPs for maintenance activities also need to include a communication strategy, including advance notice to consumers dealing with:
- potential impacts and benefits;
- shutdown notifications (if needed); and
- contacts for further information.

8.1.6 Water leakage management

Water loss is an issue for most drinking-water utilities. Water can be lost through a range of mechanisms, including leakage, unauthorized connections, metering inaccuracies and failure. Operational monitoring of water use, determination of losses and processes for identifying sources of water loss should be included in an SOP. Relative volumes of water loss should be compared with those of other utilities. If losses are higher than average or if there is evidence of increasing losses, causes should be investigated. Water leakage can lead to impacts on water quality, flows and pressure, whereas unauthorized connections can lead to contamination through poor or missing backflow prevention. Irrespective of the cause, water loss reduces potential revenue.

Guidance on developing programmes for auditing water losses and establishing management plans is available (Fanner et al., 2007; USEPA, 2010; DEWS, 2013).

8.1.7 Preventing corrosion

Corrosion can be caused by a range of factors, which vary depending on the design and construction of the distribution system (section 3.2). Operating procedures should identify key factors associated with individual systems and establish operating criteria to prevent corrosion. The primary criteria will relate to water chemistry parameters such as pH, alkalinity and age of the system. These procedures need to be linked to others, including procedures for maintaining disinfection and water flows.

Procedures for monitoring and responding to corrosion of plumbing within customers' premises may also be needed where systematic problems are caused by the nature of the drinking-water supply (e.g. where the water supply is soft and aggressive).

8.1.8 Selection of pipe materials and chemicals

All materials used in distribution systems should be suitable for contact with drinking-water. A number of countries have established testing and certification standards for such materials (Standards Australia, 2005; Japanese Standards Association, 2012; NSF International, 2012; Umweltbundesamt, 2012). Guidelines have also been established to deal with the quality and certification of chemicals used in drinking-water supplies (Drew & Frangor, 2003; NHMRC & NRMMC, 2011).

SOPs for creation of new assets, supply of materials and maintenance should be established to deal with selection of materials and chemicals, including certification requirements.

8. PREPARE MANAGEMENT PROCEDURES

8.1.9 Connection of new customers (including cross-connection control and backflow prevention)

An SOP should be established for connection of new customers. This should include requirements for installation of backflow devices. Matters that can be included in such SOPs include:
- definition of responsibilities (including legal responsibilities);
- categorization of hazards and the devices to be used;
- review and cataloguing of existing devices;
- procedures for installing devices;
- programmes for inspecting, maintaining and testing devices;
- training requirements for personnel and installers of devices (e.g. plumbers); and
- education programmes for owners/managers of buildings and facilities on the need for backflow prevention and the implications of poor control for distribution systems.

Devices can be installed on individual fixtures (e.g. cooling towers, medical or veterinary equipment), sections of water supplies within buildings or facilities (e.g. irrigation systems injected with pesticides and pesticide laboratories), property boundaries (e.g. hospitals, food processing premises) or combinations of these. The types of device can range from simple non-testable devices on low-risk premises (e.g. domestic dwellings, including dwellings with rainwater tanks) to air gaps and reduced pressure zone devices on high-risk premises (e.g. hospitals, laboratories and chemical plants). A hazard rating system should be included in the SOP to determine the type of backflow prevention device required for various types of connection. Table 15 provides examples from a rating system.

The SOP is also required to deal with inspection, maintenance and testing of backflow devices. These devices can be installed and maintained by water utilities or, alternatively, by building and facility managers. Inspection and maintenance programmes should be risk based and should include consideration of new building construction or major renovations and changes of ownership. The latter provides an opportunity for education of new owners and managers on responsibilities and risks.

In some jurisdictions, control of plumbing, including installation, inspection and maintenance of backflow prevention devices, has been transferred from water utilities to separate technical and plumbing regulators. In these cases, the water utility should ensure that SOPs are designed and implemented by the independent agency. If necessary, support for such programmes should be sought from public health agencies.

8.1.10 Repair of water main breaks

Water main breaks and leaks represent significant risks of contamination. Unlike installation of new mains, repairs are normally undertaken in wet environments with limited control of soil in the vicinity of the break, particularly in the early stages of detection and excavation. Contamination can be introduced before repair, from soil and water in trenches dug to effect repairs, from equipment and replacement pipes, fittings and materials and from incomplete repairs. SOPs are essential to minimize potential impacts on water quality.

Table 15. Typical hazard ratings for connected users

Cross-connection	Hazard rating	Backflow device
Premises		
Hospitals and clinics	High	RBT or RPZD
Chemical plants, chemical and pathology laboratories	High	RBT or RPZD
Universities	High	RBT or RPZD
Food processing plants	Medium	Testable device
Caravan parks, public swimming pools, marinas	Medium	Testable device
Domestic premises, including those with rainwater tanks	Low	Non-testable device
Sections of buildings or facilities		
Agricultural and horticultural irrigation systems injected with fertilizers, herbicides and insecticides	High	RBT or RPZD
Dockside facilities	High	RAG or RPZD
Industrial and teaching laboratories	High	RAG or RPZD
Hospital operating theatres	High	RBT or RPZD
Secondary school laboratories	Medium	Testable device
Fire services	Medium	DCV
Water filtration equipment	Low	Non-testable device
Home fire sprinklers	Low	DCV
Individual devices		
Water supply to weed and pest spraying and water cartage trucks	High	RBT or RPZD
Hospital equipment used for handling and processing chemical and microbiological processes	High	RAG or RPZD
Cooling towers	High	RAG or RPZD
Chlorinator at water treatment plant	Medium	Testable device
Haemodialysis machines	Low	Non-testable device
Drink dispensing machines, coffee machines	Low	Non-testable device

DCV: double check valve; RAG: registered air gap; RBT: registered break tank; RPZD: reduced pressure zone device

Source: Standards Australia/Standards New Zealand (2003)

SOPs should include:
- reference to accurate maps of pipework, including the location of shut-off valves and of other pipework, such as sewers;
- the location and, where practical, marking of repair equipment (i.e. distinguish equipment used to repair drinking-water mains from equipment used to repair sewers);
- guidelines for storing, inspecting and, if necessary, cleaning pipes, fittings and other materials used in repairs;
- assessing risks (e.g. vicinity of sewers) and monitoring performance of repairs;
- precautions to be applied when risks are assessed as being high;
- methods for repairing live mains (e.g. by installation of external repair clamps);
- methods for repairs involving cutting of mains;
- hygiene requirements during repairs;
- flushing and disinfection requirements before returning water mains to service;
- procedures if contamination by sewage is suspected (e.g. due to associated break of a sewer or in areas with insufficient sanitation);

- sampling requirements;
- requirements for disposing of waste materials and removing water released from bursts and repairing associated damage (e.g. to adjacent property); and
- documentation and mapping of water main breaks.

It is essential that the implementation of repairs is supervised and inspected to ensure that SOPs are applied. This is particularly important for large breaks and repairs and where risks are high – for example, due to close proximity of a sewer or in areas of poor sanitation.

Where frequent and repeated water main breaks occur or where repairs reveal that mains are in poor condition, replacement may be required. Where this involves major construction and cost, replacement should be included in the system improvement plan (section 5).

8.1.11 Construction and commissioning of new mains

Construction and commissioning of new mains or reconnecting of mains that have been out of service should be preplanned and should be undertaken in a measured manner. Relevant industry codes, standards and regulatory requirements should be identified and adopted.

A number of the hazards and procedures associated with repair of water main breaks also apply to installation of new mains, such as potential contamination from soil and water in trenches, equipment, replacement pipes, fittings and materials. SOPs should deal with these issues. In addition, SOPs should deal with separation distances from sewage lines and other buried infrastructure and materials and equipment to be used. New materials should be inspected for cleanliness and compliance with standards and certification requirements. Construction sites should be clearly marked, and barriers should be erected to prevent unauthorized access. Flushing, disinfection and testing requirements prior to commissioning should be specified. Test parameters could include *E. coli*, turbidity, disinfectant residual and pH. Any failure to meet water quality requirements should be investigated to determine whether it was caused by inadequate control of procedures during construction.

8.1.12 Dewatering and recharging distribution mains

Shutting down relatively large distribution mains (>225 mm diameter) for work or inspections can be complex, with the potential for widespread water quality impacts if not managed well. SOPs for managing these changes should include many of the procedures adopted for repairing water main breaks described above. In addition, they should consider:
- a process for assessing and managing risks to drinking-water quality, including impacts arising from flow changes in other mains;
- sanitary controls for people and equipment working near and entering the main;
- protection and inspection of open scours and manholes to prevent contamination from flooding, animals, etc.;
- procedures for recharging that prevent airlocks and creation of "white water"; and
- reinstatement procedures comparable to commissioning a new main with respect to flushing, disinfection and testing.

Following confidence in the safety of the water in the main, testing can involve taste and odour testing, particularly in cases of high water age or where there has been disturbance to pipeline and coating materials.

Usually a job-specific plan should be documented to control the sequence of hydraulic operations necessary for the shutdown and reinstatement of large mains, and this should include specific water quality controls and hold-points.

8.1.13 Permeation

Permeation through plastic pipes and gaskets has been reported (Bromhead, 1997; USEPA, 2002b). If the distribution system includes components that are susceptible to permeation by organic contaminants, such as solvents and petroleum compounds, an SOP should be established to deal with prevention of contamination. The organic compounds can be present in contaminated soils or groundwater or be introduced in spills of petroleum products.

If this type of contamination occurs, it is often detectable by consumers, as many of the compounds produce strong odours at low concentrations.

8.1.14 Collection and testing of water samples (what, where, when, how and who)

The importance of collecting samples correctly is often underestimated. Poor samples can provide misleading results. For example, samples collected from fittings that are not maintained well for sampling can include sediments, corrosion products or biofilms, whereas poor hygiene practices (associated with the sample fitting, sampling equipment or the person carrying out the sampling) can lead to contamination of samples collected for microbiological examination. Samples should be collected using standard documented procedures by personnel trained in collection, labelling, storage and transport procedures. SOPs should deal with the chain of custody from the sampling point to the analyst. SOPs should include locations and frequencies of samples collected and the test methods necessary as part of routine operational and verification monitoring programmes, as well as those collected in association with incidents and emergencies.

8.1.15 Calibrating equipment

All monitoring equipment used in distribution systems, such as on-line meters measuring disinfectant concentrations after dosing plants, field equipment used to measure disinfectant residuals, pH, turbidity and water pressure and water flow meters, need to be accurate. SOPs should be established to deal with maintenance and calibration of all equipment. The frequency of calibration will depend on the manufacturer's specifications and the criticality of the monitoring equipment, taking into account previous evidence of variable performance as a function of time. Control charts can be valuable tools for assessing variability of performance.

8.1.16 Customer complaints

Customers can be the first to detect increased turbidity, colour or off-tastes and off-odours or report illnesses linked to faults in distribution systems. SOPs should be established to receive and investigate customer inquiries and complaints. All information should be recorded and, where necessary, investigated. This is particularly important where clusters of complaints are received. Staff should be trained in how to deal with complaints and appropriate responses. These staff should be briefed in a timely manner on any known events that might lead to complaints or enquiries (e.g. routine water main flushing, tank cleaning, water main repairs). Regular reviews should be undertaken into the number of complaints, common problems, and timeliness and effect of responses.

8.2 Incident criteria and protocols

The aim is that operation of distribution systems should always comply with defined criteria and set limits; however, incidents will happen, even in the best of systems. Incident criteria are required to deal with both predictable and unforeseen types of incidents.

Some potential incidents are predictable from risk assessments and reviews of SOPs. Standard responses and corrective actions should be developed. Table 16 includes some typical examples.

8. PREPARE MANAGEMENT PROCEDURES

Incident protocols should include:
- definitions of incidents;
- operational procedures for responding to incidents;
- reporting procedures (to whom and when), including requirements for communication with external stakeholders (e.g. health agencies) and, if necessary, with the public;
- identification of roles and responsibilities for both responses and communication;
- list of contact details for key personnel and alternates;
- availability and source of emergency water supplies;
- water quality monitoring requirements;
- processes and templates for issuing public advice, such as boil water advisories;
- responsibilities for issuing notifications, such as boil water advisories (i.e. the water utility or public health agency); and
- criteria for closing incidents.

In addition to predictable events, there is always a possibility of unforeseen events occurring. While the likelihood should be low for well designed and managed systems, the consequences could be high. Therefore, incident protocols should include generic processes for dealing with unforeseen events that present a contamination risk. This should incorporate components such as identified reporting and communication requirements, established roles and responsibilities and availability of emergency supplies.

It is important to assess the effectiveness of the protocols and the readiness of organizations and personnel to respond to emergencies and incidents by conducting regular training and exercises (e.g. once per year). This is particularly important for large and unexpected events.

Incident protocols need to be regularly reviewed and kept up to date. Procedures for distributing the latest version to all relevant personnel need to be established.

All incidents should be documented and reviewed. Implications for WSPs need to be considered (see section 11).

Table 16. Examples of incident criteria and responses

Incident	Operational response
Non-compliance with water quality requirements	
• detection of *E. coli*	Should always lead to investigation of a source of potential faecal contamination and to further sampling, including collection of samples from the original location as well as potential upstream sources. Asset integrity, water pressure, disinfectant residuals and turbidity (as evidence of entry of contamination) should also be examined. Protocols could include requirements to increase disinfectant concentrations or to add additional disinfection using temporary dosing. Protocols should include procedures for water main flushing and public notification, where necessary.
• exceedance of guideline value for a health-related chemical	Most health-related chemicals require chronic exposure to cause impacts, whereby exceedance of guideline values in single samples is unlikely to cause health impacts. Responses should include confirmation by retesting, investigating potential contamination sources and assessing distribution of high chemical concentrations by testing at other locations.
Known or suspected entry of contamination	Procedures for stopping contamination, isolating impacted assets (e.g. bypassing tanks) and preventing recurrence. Protocols will also include water quality testing and water main flushing procedures.

Incident	Operational response
Water main breaks and leaks	Likely to occur reasonably frequently in moderate to large distribution systems. Repair of small breaks and leaks could be seen as a standard procedure, as they will occur in all distribution systems, and an SOP should be developed for small events (see previous section). However, incident protocols should deal with larger breaks, including those that lead to extended water supply outages. Protocols should identify when alternative supplies need to be provided. This will take into account the length of time involved and the vulnerability of consumers (e.g. aged care facilities, hospitals).
Failure of pumps and valves	Repair or replacement. Protocols should include identifying availability of spares or replacement equipment. In the case of major failures, provision of alternative supplies will need to be considered. Protocols should include procedures for preventing contamination during repairs, water main flushing and public notification, where necessary.
Loss of pressure	Examine operational control of pumps and valves, investigate possible leakage into water mains. Any faults should be repaired.
Loss of disinfectant residual at ends of distribution system	Examine: • operation of disinfectant dosing, • increased disinfectant demand in source water, • decreased water flows and demand, • operation of the system, • possible intrusion of untreated water, and • changes in water supplies (i.e. mixing). Depending on the cause, consider increasing disinfectant doses, use of booster disinfection, changes in operation of the distribution system and water main flushing.
Stagnant, dirty or contaminated water associated with resumption of an intermittent supply	Examine and isolate external sources of contamination, identify remedial action such as water main flushing.
Backflow from consumer premises	Investigate nature of contamination and source of backflow. Check installation of backflow prevention device, and check performance. Investigate installation of cross-connections or bypassing of device. Remove the source of the contamination, and, if necessary, isolate the section of the distribution system affected by the event. Test samples for presence of contaminants, flush mains and retest. If necessary, issue public advice.
Interference with infrastructure, including unauthorized access to storages	Inspect for physical evidence of interference, and test water for *E. coli* and health-related chemicals. If possible, take infrastructure out of service pending completion of testing.
Permeation by organic compounds (solvents, petroleum products)	Generally associated with plastic pipes and fittings. If detected, the source should be identified and removed. If permeation is due to soil and groundwater contamination, replacement of permeable materials may need to be considered. Pipes should be flushed and further samples collected to ensure that the contamination has been removed.

9. Develop supporting programmes

- Assemble WSP team **M1**
- Describe the water supply system **M2**
- Identify hazards & hazardous events, assess risks **M3**
- Determine and validate control measures, reassess and prioritize the risks **M4**
- Develop, implement and maintain improvement plan **M5**
- Define monitoring of control measures **M6**
- Verify the effectiveness of the WSP **M7**
- Prepare management procedures **M8**
- **Develop supporting programmes M9**
  - Conduct training of operators and employees
  - Undertake research and development to understand hazards and potential hazardous events
- Plan and carry out periodic review of the WSP **M10**
- Revise the WSP following an incident **M11**

ENABLING ENVIRONMENT

Supporting programmes are defined as activities that build capacity of employees and other stakeholders to practise good management of drinking-water supplies consistent with implementation of WSPs.

Although it is important to ensure that operators and their managers are well equipped with the knowledge to implement WSPs, it is also essential to ensure that maintenance contractors, plumbers and operators and owners of facilities connected to drinking-water suppliers are provided with sufficient knowledge to ensure that their actions are consistent with the WSP approach. This can promote the installation and maintenance of appropriate backflow prevention devices and reduce uncontrolled cross-connections.

Training of operators should start with the induction of new employees and continue through the course of employment and should include WSP implementation, hygiene procedures (personnel and equipment – e.g. not using equipment used to maintain sewerage systems for drinking-water supplies without appropriate treatment) and reporting and communication protocols. This should include the need for and importance of reporting incidents. Those dealing with consumer calls should receive training on how to respond and how to deal with specific issues and incidents.

Workforce skills qualifications framework adopted by Singapore's Public Utilities Board

The Public Utilities Board, the national water agency of Singapore, adopts the workforce skills qualifications framework to design the training programme for its employees. The workforce skills qualifications framework is a systematic approach that develops, assesses and recognizes individuals based on the industry-agreed standards. This system is governed by the local workforce development authority. Both the training institutions and trainers are required to be accredited and certified by the local workforce development authority. Through the workforce skills qualifications system, all operators are equipped with the required skills and knowledge to work within the water distribution network system, including the technical skills (operating the control systems) and soft skills (handling customers).

9. DEVELOP SUPPORTING PROGRAMMES

Training programmes on WSPs in South Africa

Water safety planning is widely practised in South Africa, and many training programmes on WSPs have been developed. Some of these programmes are listed for reference:

- Introduction to Water Safety Planning – Rand Water, South Africa (as part of induction programme for all new employees)
- Developing a Water Safety Plan – Johannesburg, Rand Water Scientific Services Water Technology Training, 21–25 September 2009. Co-organizers: Cap-Net, Rand Water, International Water Association & Global Water Operators' Partnerships Alliance (UN-HABITAT)
- Water Safety Plans for Municipal Water Supply Systems – Water Institute of Southern Africa telematic training courses – with excerpts from Water Research Commission report TT 415/09 on "The Development of a Generic Water Safety Plan for Small Community Water Supply"
- The Development of a Generic Water Safety Plan for Small Community Water Supply – Water Research Commission report TT 415/09
- Guidelines for Using the Web-enabled Water Safety Plan Tool – Water Research Commission report K5/1993. Included training workshops to introduce water safety planning, highlight key steps to be considered when developing a WSP and provide step-by-step guidance on how to use a web-enabled water safety planning tool
- Blue Water Services Audit Inspector Short-course Training Programme on Water Safety Planning – Department of Water Affairs, South Africa
- Blue Drop and No Drop Handbook – Department of Water Affairs, South Africa

Training should not be undertaken on an ad hoc basis but should be organized and documented. Ideally, training should be coordinated by a single person or group (depending on the size of the water utility) within the organization.

Other types of supporting programmes can include research and development to support better understanding of hazards and potential hazardous events and their management. Where issues have been identified, such as potential regrowth of free-living pathogens (e.g. mycobacteria, *Naegleria fowleri*, *Legionella*), research could involve determining the distribution, control and significance of pathogenic strains (e.g. links to disease). Similarly, if persistent off-odours are detected, there could be a requirement to identify sources and mechanisms for control. Causes of spikes in DBP concentrations and appearance of new DBPs could also warrant research. The significance of emerging issues could also need investigating.

Although important in well managed and established supplies, research and development should not be given priority over establishing basic and sound WSP implementation.

10. Plan and carry out periodic review of the WSP

- Assemble WSP team **M1**
- Describe the water supply system **M2**
- Identify hazards & hazardous events, assess risks **M3**
- Determine and validate control measures, reassess and prioritize the risks **M4**
- Develop, implement and maintain improvement plan **M5**
- Define monitoring of control measures **M6**
- Verify the effectiveness of the WSP **M7**
- Prepare management procedures **M8**
- Develop supporting programmes **M9**
- Plan and carry out periodic review of the WSP **M10**
- Revise the WSP following an incident **M11**

- Review WSP (by WSP team)
- Record outcome of review (by WSP team)
- Endorse changes to WSP (by management)
- Communicate changes to WSP to all relevant personnel

ENABLING ENVIRONMENT

WSPs require regular reviews to ensure that they are functioning effectively, are kept up to date and have been amended based on audits (see section 7.3) and experiences of employees, operators and managers.

Regular reviews of WSPs are necessary to ensure that changes and events that could threaten effective implementation and the distribution of safe drinking-water are regularly assessed and addressed. Changes to distribution systems, revised operational procedures, installation of new mains and equipment, connection of new consumers (particularly consumers of large volumes) and property developments can all require modifications to WSPs. Implementation can be influenced by staff changes and stakeholder changes. Incidents can also lead to revisions of WSPs (see section 11).

Although operators and managers should be engaged in preparing WSPs, there is nothing like hands-on experience for identifying faults and improved practices. Employees should be encouraged to contribute ideas for amending practices to improve operation of WSPs – for instance, during regular seminars for all relevant employees, during internal audits and via templates for reporting of new potential hazards and hazardous events.

Checklist for WSP reviews

- assessment of operation of the WSP, including implementation of:
  - previously agreed amendments
  - planned changes arising from improvement programmes
- consideration of internal and external audit reports (see section 7.3)
- consideration of incidents (see section 11)
- consideration of incidents in other drinking-water supplies
- identification of changes to distribution systems, including addition of new mains, installation of new equipment, connection of new customers and property developments
- assessment of operating procedures, including SOPs
- assessment of operational monitoring results, including compliance with operational limits
- assessment of verification results
- consideration of consumer complaints
- comparison of performance in meeting targets, guideline values, standards and regulatory requirements against historical results
- stakeholder requests

Depending on the size of the water utility, reviews of WSPs should be undertaken on a regular basis by a team including representatives from the group that originally designed the WSP (see section 1). The team should include representatives of operators with responsibility for management of distribution systems. In small organizations, reviews will be undertaken by a limited number of personnel. In large organizations, teams will be larger, and reviews could be conducted at two levels, involving operators and managers and executive management. Reviews should be conducted on an annual basis and following any major change to the distribution system.

Outcomes of reviews should be recorded, and any changes to WSPs should be endorsed by management and communicated to all relevant personnel. Recommendations for inclusions in improvement programmes should be communicated to senior management. Updated versions of the WSP should be subject to document control. It is essential that all personnel are using the appropriate version of the WSP.

11. Revise the WSP following an incident

In addition to routine reviews, WSPs should be reviewed after incidents and emergencies using a root cause analysis. Well designed and operated WSPs should reduce the likelihood of incidents; although incidents will still occur, it is important to assess whether any faults or gaps in the WSP contributed to the occurrence or severity of the incident. It is also important to assess whether the response to the incident was sufficient and effective. Every effort should be made to learn from challenges to management of drinking-water safety.

The first step following an incident is to determine the cause, identify how the incident was detected or recognized, document the response and identify impacts on drinking-water quality and supply and impacts on consumers. All records generated during the incident should be collected and reviewed.

Post-incident reviews should be comprehensive, but key issues to consider will include:
- whether the incident was caused by a new hazardous event or hazard;
- whether risks need to be reassessed where the incident was caused by a previously identified hazardous event or hazard;
- whether existing control measures are sufficient or whether upgrades or additional measures should be considered. It is important not to overreact; capital expenditure and substantial changes in practice may not be warranted if the risk of recurrence is considered to be low;
- whether operational monitoring was effective and whether it could be modified to improve timely detection of deviations from expected performance;
- whether the incident protocol functioned effectively in responding to the event and restoring normal operation of the distribution system;
- whether the documentation of the incident was sufficient to perform the post-incident review;
- whether communication, responses to consumer enquiries and issuing of public advice (if needed) were appropriate and timely;
- if emergency water supplies were required, whether they were delivered in sufficient volumes and in a timely fashion; and
- whether operators and managers responding to incidents had appropriate skills and knowledge and whether they had access to appropriate equipment and tools.

Post-incident reviews should be considered good practice and not an exercise in identifying blame or fault. The aim is to perform an open appraisal of the incident and to identify lessons learnt and opportunities to revise and improve the WSP. Recurrence of an incident due to inaction is not acceptable.

Outcomes of a root cause analysis, including recommended revisions and improvements to WSPs, should be documented. If resources permit, recommended revisions should be incorporated into WSPs immediately. Where this is not possible – for example, where significant capital expenditure is required – recommended changes and upgrades should be included in improvement plans (section 5). If the outcomes are of general interest for other water supplies, they should be published – for instance, through the national water associations.

12. Enabling environment

- Assemble WSP team **M1**
- Describe the water supply system **M2**
- Identify hazards & hazardous events, assess risks **M3**
- Determine and validate control measures, reassess and prioritize the risks **M4**
- Develop, implement and maintain improvement plan **M5**
- Define monitoring of control measures **M6**
- Verify the effectiveness of the WSP **M7**
- Prepare management procedures **M8**
- Develop supporting programmes **M9**
- Plan and carry out periodic review of the WSP **M10**
- Revise the WSP following an incident **M11**

 Regulatory and policy frameworks

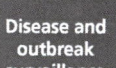 Independent surveillance, audits and inspections

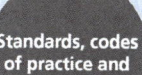 Disease and outbreak surveillance

Standards, codes of practice and certification

 Capacity building

The enabling environment refers to actions and capabilities undertaken by external stakeholders that contribute to the design of WSPs or support implementation of WSPs. These can include regulatory and policy frameworks; capacity building of operational personnel, plumbers and consumers (particularly specialist and large-volume water consumers); independent surveillance, audits and inspections; disease monitoring and outbreak detection; and activities undertaken by specialist contractors and service providers.

12.1 Regulatory and policy frameworks

Effective management of drinking-water quality should be supported by legislative frameworks. The precise nature of legislation will depend on constitutional settings, and drinking-water regulations may be developed at a national or regional level, with enforcement potentially including national, regional and local authorities. Legislation should identify how drinking-water safety can be achieved and how safety should be determined by drinking-water providers and assessed by regulators. Legislation can also deal with obligations of consumers connecting to drinking-water distribution systems. In addition to legislation, provision of safe drinking-water may be supported by standards and policies applying to matters such as certification of water operators, water treatment chemicals, distribution system materials and plumbing. These can take the form of mandatory requirements or recommended practices.

Another form of legislation deals with economic regulation, which can include price setting, customer access rights and customer service obligations.

12.1.1 Water quality legislation

Legislation should define the roles and responsibilities of the primary agencies and entities associated with ensuring the delivery of safe drinking-water. This should include regulators and drinking-water providers. Regulation of drinking-water quality requires public health expertise, which typically resides within government health agencies, although other agencies, such as those responsible for environmental protection, may play a role. In some cases, specialized government agencies may be established to administer drinking-water legislation. Local government environmental health officers may also assist in the administration of legislation. Irrespective of the identities of regulators, they need to be given sufficient power to monitor and enforce legislative requirements.

Legislation should provide a definition of a drinking-water provider, including the scope of application – that is, whether the legislation applies to government and local government utilities as well as privately owned utilities and community managed supplies. It should also provide clear direction to drinking-water providers on:
- management of water supplies, monitoring and reporting;
- relevant standards, guidelines and codes of practice; and
- how compliance will be audited, including mechanisms that will be applied by enforcement agencies.

The boundaries of legislative responsibilities differ from country to country. If legislative responsibilities include supply to the point of delivery to consumers, then all components, including installation of backflow prevention devices and plumbing requirements, could be included in a single piece of legislation. However, in many jurisdictions, the responsibility of drinking-water providers ends at the boundaries of connected properties and facilities. Separate legislative requirements can provide protection of water quality associated with connections. These requirements can deal with installation of backflow prevention devices, plumbing standards, including installation, certification and general responsibilities to prevent contamination of the distribution network.

South Australian and Victorian Safe Drinking Water Acts

The *South Australian Safe Drinking Water Act 2011* and the *Victorian Safe Drinking Water Act 2003* were developed to protect public health, to apply a consistent approach to urban and rural drinking-water suppliers and to provide certainty and direction to drinking-water suppliers on how to achieve safe drinking-water and how compliance can be measured.

Both Acts are administered by the state health departments and apply to public drinking-water supplies. Responsibilities of both are identified. The Acts are based on implementing the principles incorporated in the 2011 Australian Drinking Water Guidelines and include a focus on developing and implementing risk management plans (equivalent to WSPs).

In addition to risk management plans, the South Australian and Victorian Safe Drinking Water Acts share a number of other common features, including:
- requirements for auditing of risk management plans by independent auditors;
- reporting of known or suspected incidents that could lead to unsafe drinking-water;
- reporting of results to the health department;
- reporting of results to the public;
- approval of testing laboratories;
- approval of auditors;
- appointment of authorized officers who can inspect water supplies and investigate complaints; and
- powers to take action, require remedial action and, if necessary, issue public notices (e.g. boil water advisories).

Guidance is provided in the Acts and Regulations on how various requirements are to be met. For example, directions are provided on information that should be included in risk management plans. Additional guidance is also provided on dedicated websites: http://www.sahealth.sa.gov.au/safedrinkingwateract and http://www.health.vic.gov.au/water/drinkingwater/

Source: Based on the *South Australian Safe Drinking Water Act 2011* and the *Victorian Safe Drinking Water Act 2003* and supporting regulations

12.1.2 Economic legislation

Economic regulators usually act independently from the government and aim to provide consumers with value for money by establishing the limit on how much individual water utilities can charge their customers through a process known as the price review. In addition, there are also independent bodies that act on behalf of the customers and help to protect the standard of service customers receive from water utilities. Customers who are unhappy about the service they have received from their water utility can contact such an independent body for advice about their complaints, which typically include the number of unplanned interruptions, properties at risk of low pressure and the percentage of complaints that were not answered within 5 days.

12.2 Independent surveillance, audits and inspections

The final element in the WHO framework for safe drinking-water is independent surveillance by agencies responsible for public health. In most cases, surveillance of distribution systems will be included as a component of surveillance of the entire drinking-water supply.

Surveillance is an essential activity to assess and review the safety and acceptability of drinking-water supplies. It should be coordinated by the drinking-water regulator or, in the absence of legislation, the public health agency. Surveillance can involve audits, direct assessment by testing or a combination of both. Audits will involve reviewing the design and implementation of WSPs, including an assessment of monitoring results. Direct assessment requires that the surveillance agency has the expertise to collect appropriate samples and has access to suitable and preferably accredited laboratories. The role of audits is to determine the capability to consistently produce safe drinking-water, whereas direct assessment determines whether safe drinking-water was produced at the time of testing.

When designing surveillance programmes, consideration should be given to whether audits will be the sole responsibility of public health agencies or will include audits undertaken by third parties (e.g. specialist auditors). Where third parties are used, the public health agency needs to retain overall responsibility for surveillance. This should include determining the frequency of audits, procedures to be applied and reporting requirements. Frequencies should be based on size, complexity and risk associated with individual drinking-water supplies. Additional audits could be required following substantial changes to distribution systems or following significant incidents.

12.3 Disease and outbreak surveillance

Monitoring of disease trends can provide evidence of the need to improve the management of water distribution systems. Once a new strategy has been implemented, information on disease trends can provide evidence of the strategy's impact. Public health agencies should establish networks with professional bodies to help detect disease and to disseminate public health information. They need to establish criteria that would initiate an investigation and procedures on how such investigations should be performed. This would include procedures for identifying and confirming potential sources of disease.

12.3.1 Disease surveillance

The aim of WSPs is to reduce the likelihood of waterborne disease and outbreaks. In the longer term, disease surveillance can provide evidence to support the effectiveness of WSPs. In the shorter term, detection of disease has led to the subsequent identification of faults in distribution systems.

Surveillance can take a few forms, including:
- a structured systematic process, generally coordinated by a central health agency based on mandatory reporting by medical practitioners and clinical laboratories of specified diseases (e.g. cholera, cryptosporidiosis, waterborne Legionnaires' disease);
- a structured process coordinated by a central health agency based on voluntary reporting by jurisdictional health agencies; and
- normal work practices for agencies with environmental health responsibilities where reports of disease are received by the agency from a number of sources and investigated.

The first type of surveillance is typically focused on microbial disease, whereas the latter two types can detect disease caused by microbial and chemical contamination.

12. ENABLING ENVIRONMENT

Many countries have established programmes for surveillance and reporting of communicable diseases. These programmes tend to focus on mandatory reporting of specific pathogens and conditions rather than the source of those organisms. There are a number of objectives of this type of surveillance, including early detection and response to outbreaks, monitoring longer-term disease trends and assessing impacts of interventions and control programmes. The latter could include specific interventions, such as the Dracunculus Eradication Programme, or control measures, such as improved management of drinking-water supplies through application of WSPs. Although outbreaks associated with distribution systems tend to be smaller in size, the localized nature of these outbreaks may aid detection.

Detection of disease associated with chemicals is particularly significant for distribution systems, as they are more common causes of illness compared with other types of waterborne outbreaks. In the USA between 1971 and 2002, chemicals were associated with 35% of distribution system outbreaks, compared with 11% for all outbreaks (Craun et al., 2006).

Acute chemical poisoning can have a rapid onset, leading to immediate investigation and, in some cases, identification of distribution system faults.

12.3.2 Disease and outbreak investigation

Single cases of acute illness from chemical exposure will usually lead to an investigation with a reasonable likelihood of identifying the cause. If the source of the chemical is contamination of a distribution system, identification of the source and prevention of further cases need to be undertaken immediately.

Investigations of microbial disease tend to focus on outbreaks except for specific circumstances, such as cases of amoebic meningitis caused by *Naegleria fowleri* or unusual occurrence of a disease (e.g. in locations where the disease is rare). Outbreaks are generally defined as two or more cases linked in location and time and should be investigated by established protocols to confirm occurrence, followed by identification of the source. Outbreak investigations should follow a planned set of activities, including:
- establishing responsibility/leadership;
- outbreak confirmation, including case definition incorporating descriptions of location, time of onset, personal characteristics of cases (e.g. number of cases, age, sex, travel and other movements), and biological and clinical criteria (symptoms and test results). As they arise, cases should be categorized as definite, probable or possible, based on the level of data available;
- case mapping (locations and time);
- hypothesis generation and confirmation. In the case of waterborne disease, confirmation will usually involve collecting and analysing water samples, identifying the fault and successful implementation of remedial action; and
- communication of findings and reassurance of containment and future prevention.

Waterborne disease outbreaks and cases of chemical-related illness or disease associated with distribution systems represent preventable failures of WSPs. Priorities are to interrupt/stop transmission, minimize the magnitude of the outbreak and prevent further cases; identify the fault or cause of the disease or illness; repair the fault and prevent recurrence; and, finally, implement appropriate changes to WSPs.

Disease investigation will be led by public health agencies, but expert input from water utilities will be required where drinking-water is a potential source.

12.3.3 Communication, recording and reporting

Effective disease surveillance is underpinned by comprehensive and practical reporting and recording systems. These systems need to include case definitions, notification and reporting requirements, data management, evaluation, outbreak preparedness and training. Reporting requirements for medical practitioners and laboratories should include directions on information needs to assist in identifying sources of infection.

Communication between water utilities and surveillance agencies is essential to improve understanding of functions and possible links between drinking-water distribution and disease. Water utilities need to communicate significant distribution faults and incidents to surveillance agencies, whereas surveillance agencies need to communicate information about potential waterborne outbreaks to water utilities. Investigation of this type of outbreak will require close cooperation between the surveillance agency and water utility. As a first step, comparison of the geographical distribution of cases with physical features of distribution system networks can quickly confirm or rule out links to drinking-water supplies. Networks need to be established to support rapid and direct communication, which are essential for effective investigation.

12.3.4 Lessons learnt

It is important to record and report outcomes of disease and outbreak investigations. On a regional, national and possibly international level, communicating the lessons learnt, including sharing experiences in detecting disease, identifying deficiencies and implementing responses, can reduce the likelihood of occurrence in other systems.

12.4 Standards, codes of practice and certification

12.4.1 Distribution system design codes

Regulatory agencies and water utility associations can issue codes and guideline documents dealing with design and construction requirements for distribution systems (Water Services Association of Australia, 2002; Ontario Ministry of the Environment, 2008; Great Lakes – Upper Mississippi River Board of State and Provincial Public Health and Environmental Managers, 2012). These codes and guidelines can provide direction and guidance on a broad range of issues, including:

- system planning,
- hydraulic design,
- structural design,
- water storages,
- pumping facilities,
- system construction,
- products and materials,
- separation distances from contamination sources and other pipework,
- backflow and cross-connection control, and
- pipe flushing and swabbing and disinfection.

Codes have also been developed for dual reticulation systems involving supply of drinking-water and non-drinking-water supplies to premises (Water Services Association of Australia, 2005).

Whereas guidelines and codes typically provide technical specifications and descriptions of good practice to be adopted on a voluntary basis, they can be adopted as default requirements by independent technical regulators and certifiers of distribution systems.

12.4.2 National standards and certification systems

Devices and materials used in water distribution systems need to meet quality requirements and comply with applicable standards and codes of practice. Some countries have established standard-setting bodies and certification systems to provide assurance that, when used in accordance with design specifications, devices and materials will perform as required and be safe. Standards and codes of practice can apply to a range of activities associated with the design, installation and management of distribution systems, including:

- selection of materials used in contact with drinking-water both in distribution systems and within buildings. Material standards can deal with physical attributes and ensure that products do not give rise to unacceptable contamination of water or support microbial growth;
- building and plumbing codes that describe installation requirements within premises, including backflow prevention device specifications and selection criteria;
- installation of secondary or alternative water supplies (e.g. recycled water and rainwater with connections to drinking-water systems); and
- water sampling and testing methods.

Standards are typically developed in cooperation with manufacturers, technical experts, regulatory agencies, certifying agencies and consumers. Public health agencies should participate in developing or approving parts of standards that are intended to protect public health. Standards can:

- represent technical provisions and norms to be adopted on a voluntary basis as good practice;
- be adopted as requirements by government or local government authorities; and
- be adopted by reference in regulations.

Standards should include criteria for achieving, measuring and certifying compliance. Certification is used to confirm that devices and materials used in water systems meet standards or alternative criteria. Certification can be undertaken by government agencies or private organizations. Certification agencies may assess data and information provided by manufacturers, undertake specific testing or conduct inspections and audits. Certification may be issued subject to application of defined conditions. These conditions could identify specific applications and uses of certified products (e.g. where devices can and cannot be used).

Independent inspection and certification of building and facility plumbing systems prior to connection with distribution systems provide assurance that design principles have been applied and that appropriate barriers to contamination have been included.

Standard setting and certification also apply to sample collection and laboratory analysis. Samples need to be collected, stored and transported using established procedures and appropriate equipment (e.g. correctly prepared sample bottles). Similarly, laboratories need to be competent to perform the tests that they undertake. This includes using suitable methods, appropriate testing equipment, and qualified and capable personnel. Some countries have established standards supported by certification and accreditation systems for laboratory services.

Accreditation of testing laboratories provides a level of quality assurance and control that gives confidence to drinking-water providers and regulators in the accuracy of test results. Both false negatives and false positives can have far-reaching effects in either missing potentially dangerous situations or leading to expensive remedial action and potentially issuing of public warnings when not required.

12.5 Capacity building

12.5.1 Training

Design, installation and management of water distribution systems can involve a range of personnel, all of whom must be competent to undertake assigned or required tasks. This involves education, training and, in some cases, certification and registration.

Certification of water system operators supports good practice and application of WSPs. Certification could include a series of levels through which operators can progress and should preferably include consideration of both training and experience. Responsibilities should be aligned with certification levels.

Plumbing is recognized as a technical trade (WHO & WPC, 2006). In many countries, plumbers have to be registered to be able to practise. Registration follows successful completion of apprenticeships with registered plumbers or college-based courses. Similar to water system operators, registration may include more than one level, based on skills and years of experience. Where registration systems exist, there should also be powers to suspend or revoke registration due to substandard work. The capacity of plumbers should be supported by ongoing education and issuing of codes of practice and guidelines. These can be provided by national or regional plumbing industry associations and by drinking-water providers and associations.

Education and guidance should be provided to owners, managers and operators of connected buildings and facilities. In the case of specialist and large-volume consumers, this could include guidance on requirements to prevent cross-connections and backflows from storages and water-based devices (e.g. cooling towers, boilers). Education of owners of domestic and residential dwellings should include information on requirements relating to installation or modification of plumbing (e.g. whether a registered plumber is required) and requirements relating to cross-connection with alternative supplies (e.g. rainwater tanks, recycled water systems where dual reticulation supplies are installed).

12.5.2 Training providers

Training providers can provide courses to support competence. In some cases, course work can be combined with supervised "on-the-job" training. Training should be consistent with existing regulations, standards, codes of practice and requirements of regulatory authorities.

Training can be provided by water companies, institutes and professional associations. In some countries, training programmes are subject to certification and accreditation programmes. The aim of training programmes is to produce personnel with sufficient expertise and training to undertake specific tasks.

12.5.3 Maintenance contractors

Maintenance contractors may be used by water utilities to undertake specific tasks associated with the management of distribution systems. Water utilities should only engage contractors who can demonstrate competence and compliance with relevant formal requirements (e.g. certification). Service providers need to be able to demonstrate competence in undertaking tasks for which they contract. In some cases, certification programmes have been established. In other cases, levels of service or training may be specified by industry associations. Service providers need to be able to provide evidence of compliance with established programmes and, where applicable, certification. Service providers should provide evidence in the form of formal reports or certificates of completion to demonstrate that tasks have been completed in accordance with requirements.

12. ENABLING ENVIRONMENT

12.5.4 Independent auditors

Some jurisdictions use and certify independent auditors to determine the effectiveness of WSPs and compliance with occupational health and safety requirements. Qualifications or auditors should incorporate a combination of knowledge and experience. Auditors should have expertise in assessing documentation and reporting mechanisms. Auditors may be required to submit reports on their findings to the regulatory agency.

12.5.5 Risk assessors

Risk assessors need to have the expertise, knowledge and resources to undertake the task competently. Risk assessors should have expertise in:
- public health aspects of water quality;
- local legislative requirements, standards and codes of practice;
- development of WSPs;
- water distribution systems;
- identification of hazards and potential sources of these hazards;
- determination of risk;
- identification and assessment of appropriate control measures;
- operational monitoring procedures to ensure that the control measures remain effective; and
- verification procedures.

Risk assessors need to comply with formal requirements, including certification and approval conditions established by regulatory agencies. If unacceptable risks are identified, they should be reported immediately to whomever commissioned the assessment. If a serious and potentially immediate risk to public health is identified, notification of the regulatory authority will be required.

References

ABNT (2009). ABNT/NBR 15784/2009: Produtos Químicos, utilizados no tratamento de água para consumo humano – Efeitos à Saúde – Requisitos. Associação Brasileira de Normas Técnicas.

Acker C, Kraska D (2001). Operation and maintenance of water distribution systems during and after conversion to chloramine disinfection. In: Proceedings of the AWWA Annual Conference. Denver (CO): American Water Works Association.

Ailes E, Budge P, Shankar M, Collier S, Brinton W, Cronquist A et al. (2013). Economic and health impacts associated with a *Salmonella* Typhimurium drinking water outbreak – Alamosa, CO, 2008. PLoS One. 8(3):e57439.

Ainsworth R, editor (2004). Safe piped water: managing microbial water quality in piped distribution systems. Geneva: World Health Organization (http://www.who.int/water_sanitation_health/dwq/924156251X/en/, accessed 6 May 2014).

AWWA (American Water Works Association) (1992). Water Industry Database: utility profiles. Denver (CO): American Water Works Association.

AWWA (American Water Works Association) (2004). Guidelines for the physical security of water utilities. Denver (CO): American Water Works Association.

AWWA (American Water Works Association) (2008). AWWA/ANSI C104: Cement mortar lining for ductile iron pipe and fittings. Denver (CO): American Water Works Association.

Bartram J, Cotruvo J, Exner M, Fricker C, Glasmacher A, editors (2003). Heterotrophic plate counts and drinking-water safety: the significance of HPCs for water quality and human health. Geneva: World Health Organization (http://www.who.int/water_sanitation_health/dwq/hpc/en/, accessed 6 May 2014).

Bartram J, Corrales L, Davison A, Deere D, Drury D, Gordon B et al. (2009). Water safety plan manual: step-by-step risk management for drinking-water suppliers. Geneva: World Health Organization (http://www.who.int/water_sanitation_health/publication_9789241562638/en/, accessed 6 May 2014).

Barwick RS, Levy DA, Craun GF, Beach MJ, Calderon RL (2000). Surveillance for waterborne-disease outbreaks – United States, 1997–1998. MMWR CDC Surveill Summ. 49(4):1–35.

Benjamin M, Sontheimer H, Leroy P (1996). Corrosion of iron and steel. In: Internal corrosion of water distribution systems, second edition. Denver (CO): American Water Works Association Research Foundation; and Karlsruhe: DVGW-Technologiezentrum Wasser.

Berend K, Trouwborst T (1999). Cement-mortar pipes as a source of aluminum. J Am Water Works Assoc. 99(7):91–100.

Blackburn BG, Craun GF, Yoder JS, Hill VH, Calderon RL, Chen N et al. (2004). Surveillance for waterborne-disease outbreaks associated with drinking water – United States, 2001–2002. MMWR CDC Surveill Summ. 53(8):23–45.

Boffardi BP (1988). Lead in drinking water – causes and cures. Public Works. 119(11):67–70.

Bonds RW (2005). Cement mortar linings for ductile iron pipe. Birmingham (AL): Ductile Iron Pipe Research Association.

REFERENCES

Boulay N, Edwards M (2001). Role of temperature, chlorine, and organic matter in copper corrosion byproduct release in soft water. Water Res. 35(3):683–90.

British Standards Institution (2000). BS6920-2.2.1:2000: Suitability of non-metallic products for use in contact with water intended for human consumption with regard to their effect on the quality of water. London: British Standards Institution.

Bromhead J (1997). Permeation of benzene, trichloroethene and tetrachloroethene through plastic pipes: an assessment for the Drinking Water Inspectorate. London: Drinking Water Inspectorate (http://dwi.defra.gov.uk/research/completed-research/reports/dwi0772.pdf, accessed 6 May 2014).

Broo AE, Berghult B, Hedberg T (2001). Drinking water distribution – Improvements of the surface complexation model for iron corrosion. Water Sci Technol Water Supply. 1(3):11–8.

Brunkard JM, Ailes E, Roberts VA, Hill V, Hilborn ED, Craun GF et al. (2011). Surveillance for waterborne-disease outbreaks associated with drinking water – United States, 2007–2008. MMWR CDC Surveill Summ. 60(12):38–74.

Campbell R (2011). Smaller New Orleans after Katrina, census shows. New York Times, 3 February 2011.

Cantor AF, Park JK, Vaiyavatjamai P (2000). The effect of chlorine on corrosion in drinking water systems. Final report. Midwest Technology Assistance Center, University of Illinois and Illinois State Water Survey (http://mtac.isws.illinois.edu/mtacdocs/CorrosionFinRpt/CorrosnFnlRpt00.pdf, accessed June 2013).

CDC (Centers for Disease Control and Prevention) (1981). Water-related diseases surveillance annual summary 1981. Atlanta (GA): Centers for Disease Control and Prevention (http://www.cdc.gov/healthywater/pdf/mmwr/MMWR_WBDOSS_Summary_1981.pdf, accessed 7 May 2014).

CDC (Centers for Disease Control and Prevention) (1982). Water-related diseases surveillance annual summary 1982. Atlanta (GA): Centers for Disease Control and Prevention (http://www.cdc.gov/healthywater/pdf/mmwr/MMWR_WBDOSS_Summary_1982.pdf, accessed 7 May 2014).

CDC (Centers for Disease Control and Prevention) (1983). Water-related diseases surveillance annual summary 1983. Atlanta (GA): Centers for Disease Control and Prevention (http://www.cdc.gov/healthywater/pdf/mmwr/MMWR_WBDOSS_Summary_1983.pdf, accessed 7 May 2014).

CDC (Centers for Disease Control and Prevention) (1984). Water-related diseases surveillance annual summary 1984. Atlanta (GA): Centers for Disease Control and Prevention (http://www.cdc.gov/healthywater/pdf/mmwr/MMWR_WBDOSS_Summary_1984.pdf, accessed 7 May 2014).

CDC (Centers for Disease Control and Prevention) (2011). Legionellosis – United States, 2000–2009. MMWR Morb Mortal Wkly Rep. 60(32):1083–6.

CDC (Centers for Disease Control and Prevention) (2014). *Naegleria fowleri* – Primary amebic meningoencephalitis (PAM): public drinking water systems. Atlanta (GA): Centers for Disease Control and Prevention (http://www.cdc.gov/parasites/naegleria/public-water-systems.html#r5, accessed 7 May 2014).

Christy P, Robinson B (1984). Disinfection of water for control of amoebae. Water. 11:21–4.

Clark RM (1995). Modelling water quality changes in distribution systems: a U.S. perspective. In: Cabrera E, Vela AF, editors. Improving efficiency and reliability of water distribution systems. Water Science and Technology Library; 395–414.

Colorado Department of Public Health and Environment (2009). Waterborne *Salmonella* outbreak in Alamosa, Colorado, March and April 2008: outbreak identification, response and investigation. Denver (CO): Colorado Department of Public Health and Environment, Water Quality Control Division, Safe Drinking Water Program (http://www.colorado.gov/cs/Satellite?blobcol=urldata&blobheadername1=Content-Disposition&blobheadername2=Content-Type&blobheadervalue1=inline%3B+filename%3D%22Alamosa+Outbreak+Investigation+Report+.pdf%22&blobheadervalue2=application%2Fpdf&blobkey=id&blobtable=MungoBlobs&blobwhere=1251807327157&ssbinary=true, accessed May 2014).

Conroy PJ, Kings K, Olliffe T, Kennedy G, Blois S (1994). Durability and environmental impact of cement mortar linings. Swindon, Wiltshire: Water Research Centre (Report No. FR 0473).

Co-operative Research Centre for Water Quality and Treatment (2007a). Health Stream, Issue 47 (http://www.waterra.com.au/publications/health-stream/, accessed January 2014).

Co-operative Research Centre for Water Quality and Treatment (2007b). Health Stream, Issue 48 (http://www.waterra.com.au/publications/health-stream/, accessed January 2014).

Craun GF (2012). The importance of waterborne disease outbreak surveillance in the United States. Ann Ist Super Sanita. 48(4):447–59.

Craun GF, Calderon RL (2001). Waterborne disease outbreaks caused by distribution system deficiencies. J Am Water Works Assoc. 93(9):64–75.

Craun MF, Craun GF, Calderon RL, Beach ML (2006). Waterborne outbreaks reported in the United States. J Water Health. 4(Suppl 2):19–30.

Cunliffe DA (1991). Bacterial nitrification in chloraminated water supplies. Appl Environ Microbiol. 57(11):3399–3402.

Cunliffe D, Bartram J, Briand E, Chartier Y, Colbourne J, Drury D et al., editors (2011). Water safety in buildings. Geneva: World Health Organization (http://www.who.int/water_sanitation_health/publications/2011/9789241548106/en/, accessed 6 May 2014).

Declerck P, Behets J, Margineanu J, Hoef V, De Keersmaecker B, Ollevier F (2009). Replication of *Legionella pneumophila* in biofilms of water distribution pipes. Microbiol Res. 164:593–603.

DEWS (Department of Energy and Water Supply) (2013). Guidelines for the preparation of a System Leakage Management Plan. State of Queensland, Australia (http://www.dews.qld.gov.au/__data/assets/pdf_file/0006/78099/preparing-slmp.pdf, accessed May 2014).

Dodrill DM, Edwards M (1995). Corrosion control on the basis of utility experience. J Am Water Works Assoc. 87(7):74–85.

Dorsch MM, Cameron AS, Robinson BS (1983). The epidemiology and control of primary amoebic meningoencephalitis with particular reference to South Australia. Trans R Soc Trop Med Hyg. 77(3):372–7.

Douglas BD, Merrill DT (1991). Control of water quality deterioration caused by corrosion of cement-mortar pipe linings. Denver (CO): American Water Works Association Research Foundation and American Water Works Association.

Douglas BD, Merrill DT, Catlin JO (1996). Water quality deterioration from corrosion of cement-mortar linings. J Am Water Works Assoc. 88(7):99.

REFERENCES

Drew R, Frangor J (2003). Overview of national and international guidelines and recommendations on the assessment and approval of chemicals used in the treatment of drinking water. Canberra: National Health and Medical Research Council, Drinking Water Treatment Chemicals Working Party (http://www.nhmrc.gov.au/_files_nhmrc/publications/attachments/watergde.pdf, accessed 6 May 2014).

Du H, Li J, Moe B, McGuigan CF, Shen S, Li XF (2013). Cytotoxicity and oxidative damage induced by halobenzoquinones to T24 bladder cancer cells. Environ Sci Technol. 47(6):2823–30.

DVGW (2011). DVGW Standard W 270: Microbiological enhancement on materials in contact with drinking water – testing and assessment. Bonn: DVGW CERT GmbH.

Edwards M, Ferguson JF (1993). Accelerated testing of copper corrosion. J Am Water Works Assoc. 85(10):105–13.

Edwards M, Triantafyllidou S (2007). Chloride-to-sulfate mass ratio and lead leaching to water. J Am Water Works Assoc. 99(7):96–109.

Edwards M, Schock MR, Meyer TE (1996). Alkalinity, pH, and copper corrosion by-product release. J Am Water Works Assoc. 88(3):81–94.

Edwards M, Jacobs S, Dodrill DM (1999). Desktop guidance for mitigating Pb and Cu corrosion by-products. J Am Water Works Assoc. 91(5):66–77.

Ercumen A, Gruber JS, Colford JM (2014). Water distribution system deficiencies and gastrointestinal illness: a systematic review and meta-analysis. Environ Health Perspect. 122(7):651–60.

European Committee for Standardization (2014). Home page (https://www.cen.eu, accessed 6 May 2014).

Fanner PV, Sturm R, Thornton J, Liemberger R, Davis SE, Hoogerwef T (2007). Leakage management technologies. Denver (CO): American Water Works Association Research Foundation.

Flannery B, Gelling LB, Vugia DJ, Weintraub JM, Salemo JJ, Conroy MJ (2006). Reducing *Legionella* colonization of water systems with monochloramine. Emerg Infect Dis. 12(4):588–96.

Flournoy RL, Monroe D, Chestnut N-H, Kumar V (1999). Health effects from vinyl chloride monomer leaching from pre-1977 PVC pipe. In: Proceedings of the AWWA Annual Conference. Denver (CO): American Water Works Association.

Folkman S (2012). Water main break rates in the USA and Canada: a comprehensive study. Logan (UT): Utah State University, Buried Structures Laboratory (http://www.neng.usu.edu/mae/faculty/stevef/UtahStateWaterBreakRatesLR.pdf, accessed 6 May 2014).

Frost FJ, Craun G, Calderon RL (1996). Waterborne disease surveillance. J Am Water Works Assoc. 88(9):66–75.

Glaza EC, Park JK (1992). Permeation of organic contaminants through gasketed pipe joints. J Am Water Works Assoc. 84(7):92–100.

Gou Q, Toomuluri PJ, Eckert JO (1998). Leachability of regulated metals from cement-mortar linings. J Am Water Works Assoc. 90(3):62–73.

Great Lakes – Upper Mississippi River Board of State and Provincial Public Health and Environmental Managers (2012). Recommended standards for water works. Policies for the review and approval of plans and specifications for public water supplies. Albany (NY): Health Research Inc. (http://10statesstandards.com/waterrev2012.pdf, accessed 4 May 2014).

Health Canada (2009). Guidance on controlling corrosion in drinking water distribution systems. Ottawa: Health Canada (http://www.hc-sc.gc.ca/ewh-semt/pubs/water-eau/corrosion/index-eng.php, accessed 4 May 2014).

Herwaldt BL, Craun GF, Stokes SL, Juranek DD (1991). Waterborne-disease outbreaks, 1989–1990. MMWR CDC Surveill Summ. 40(3):1–22.

Hilborn ED, Wade TJ, Hicks L, Garrison L, Carpenter J, Adam E et al. (2013). Surveillance for waterborne disease outbreaks associated with drinking water and other nonrecreational water – United States, 2009–2010. Morb Mortal Wkly Rep. 62(35):714–20.

Horsley MB, Northup BW, O'Brien WJ, Harms LL (1998). Minimizing iron corrosion in lime softened water. In: Proceedings of the 1998 AWWA Water Quality Technology Conference, San Diego, CA. Denver (CO): American Water Works Association.

Hrudey SE, Hrudey EJ (2004). Safe drinking water: lessons from recent outbreaks in affluent countries. London: IWA Publishing.

Hunter PR, Chalmers RM, Hughes S, Syed Q (2005). Self-reported diarrhea in a control group: a strong association with reporting of low-pressure events in tap water. Clin Infect Dis. 40:32–4.

ISO/IEC (International Organization for Standardization/International Electrochemical Commission) (2013). International Standard ISO/IEC 17067: Conformity assessment – Fundamentals of product certification and guidelines for product certification schemes. Geneva: International Organization for Standardization (http://www.iec-ilac-iaf.org/meetings/IEC_ILAC_IAF_3037_DEC.pdf, accessed 6 May 2014).

Japanese Standards Association (2012). Japanese Industrial Standard JIS S 3200: Equipment for water supply service. Tokyo: Japanese Standards Association.

Jones IG, Roworth M (1996). An outbreak of *Escherichia coli* O157 and campylobacteriosis associated with contamination of a drinking water supply. Public Health. 110:277–82.

Karalekas PC, Ryan CR, Taylor FB (1983). Control of lead, copper, and iron pipe corrosion in Boston. J Am Water Works Assoc. 75(2):92–5.

Kirmeyer GJ, Friedman M, Martel K, Howie D, LeChevallier M, Abbaszadegan M et al. (2001). Pathogen intrusion into the distribution system. Denver (CO): American Water Works Association Research Foundation.

Kirmeyer GJ, LeChevallier M, Barbeau H, Martel K, Thompson G, Radder L et al. (2004). Optimizing chloramine treatment. Project 2760. Denver (CO): American Water Works Association Research Foundation and American Water Works Association (AwwaRF Report 90993).

Kool J, Carpenter J, Fields B (1999). Effect of monochloramine disinfection of municipal drinking water on risk of nosocomial Legionnaires' disease. Lancet. 353:272–7.

Kramer MH, Herwaldt BL, Calderon RL, Juranek DD (1996). Surveillance for waterborne-disease outbreaks – United States, 1993–1994. MMWR CDC Surveill Summ. 45(1):1–33.

Krasner SW (2009). The formation and control of emerging disinfection by-products of health concern. Phil Trans A Math Phys Eng Sci. 367(1904):4077–95.

Krasner SW, Amy G (1995). Jar-test evaluations of enhanced coagulation. J Am Water Works Assoc. 87(10):93.

Krasner SW, Mitch WA, Westerhoff P, Dotson A (2012). Formation and control of emerging C- and N-DBPs in drinking water. J Am Water Works Assoc. 104(11):E582–95.

LeChevallier MW, Gullick RW, Karim MR, Friedman M, Funk JE (2003). The potential for health risks from intrusion of contaminants into the distribution system from pressure transients. J Water Health. 1:3–14.

Lee RG, Becker WC, Collins DW (1989). Lead at the tap: sources and control. J Am Water Works Assoc. 81(7):52–62.

Lee SH, Levy DA, Craun GF, Beach MJ, Calderon RL (2002). Surveillance for waterborne-disease outbreaks – United States, 1999–2000. MMWR CDC Surveill Summ. 51(8):1–45.

LeRoy P, Schock MR, Wagner I, Holtschulte H (1996). Cement-based materials. In: Internal corrosion of water distribution systems. Denver (CO): American Water Works Association Research Foundation and American Water Works Association.

Levine WC, Stephenson WT, Craun GF (1990). Waterborne disease outbreaks, 1986–1988. MMWR CDC Surveill Summ. 39(1):1–9.

Levy DA, Bens MS, Craun GF, Calderon RL, Herwaldt BL (1998). Surveillance for waterborne-disease outbreaks – United States, 1995–1996. MMWR CDC Surveill Summ. 47(5):1–34.

Liang JL, Dziuban EJ, Craun GF, Hill V, Moore MR, Gelting RJ et al. (2006). Surveillance for waterborne-disease outbreaks associated with drinking water – United States, 2003–2004. MMWR CDC Surveill Summ. 55(12):31–65.

Lytle DA, Schock MR (1996). Stagnation, time, composition, pH and orthophosphate effects on metal leaching from brass. Cincinnati (OH): United States Environmental Protection Agency (EPA/600/R-96-103).

Lytle DA, Schock MR (2000). Impact of stagnation time on metal dissolution from plumbing materials in drinking water. J Water Supply Res Technol – Aqua. 49(5):243–57.

Lytle DA, Sorg TJ, Frietch C (2004). Accumulation of arsenic in drinking water distribution systems. Environ Sci Technol. 38(20):5365–72.

Macharia LN (2012). Water safety in informal settlements. Presentation at the IWA Water Safety Conference, Kampala, Uganda (http://www.wsportal.org/uploads/IWA%20Toolboxes/WSP/WS%20conference/volta/LNjambi.pdf, accessed 6 May 2014).

Mangas S, Fitzgerald DJ (2003). Exposures to lead require ongoing vigilance. Bull World Health Organ. 81(11):847.

Mermin JH, Villar R, Carpenter J, Roberts L, Samaridden A, Gasanova L et al. (1999). A massive epidemic of multidrug-resistant typhoid fever in Tajikistan associated with consumption of municipal water. J Infect Dis. 179:1416–22.

Moore AC, Herwaldt BL, Craun GF, Calderon RL, Highsmith AK, Juranek DD (1993). Surveillance for waterborne-disease outbreaks – United States, 1991–1992. MMWR CDC Surveill Summ. 42(5):1–22.

Morran J, Whittle M, Leach J, Harris M (2009). A new source of NDMA in potable water supplies. Water. 36(5):99–101.

Morran J, Whittle M, Fabris RB, Harris M, Leach JS, Newcombe G et al. (2011). Nitrosamines from pipeline materials in drinking water distribution systems. J Am Water Works Assoc. 103(10):76–83.

Nairobi City and Water Sewerage Company and Athi Water Services Board (2009). Strategic guidelines for improving water and sanitation services in Nairobi's informal settlements. Nairobi: Nairobi City and Water Sewerage Company and Athi Water Services Board (http://www.wsp.org/UserFiles/file/Af_Nairobi_Strategic_Guidelines.pdf, accessed 6 May 2014).

National Research Council (2005). Public water supply distribution systems: assessing and reducing risks – First report. Washington (DC): The National Academies Press (http://www.nap.edu/catalog/11262.html, accessed 6 May 2014).

National Research Council (2006). Drinking water distribution systems: assessing and reducing risks. Washington (DC): The National Academies Press (http://www.nap.edu/catalog.php?record_id=11728, accessed 6 May 2014).

National Water Commission (2013). National performance report 2011–2012: urban water utilities. Canberra: National Water Commission.

NHMRC (National Health and Medical Research Council), NRMMC (Natural Resources Management Ministerial Council) (2011). Australian drinking water guidelines. Canberra: National Health and Medical Research Council (http://www.nhmrc.gov.au/guidelines/publications/eh52, accessed 6 May 2014).

NSF International (2012). NSF/ANSI Standard 61: Drinking water system components – Health effects. Ann Arbor (MI): NSF International.

NSF International (2013a). NSF/ANSI Standard 60: Drinking water treatment chemicals – Health effects. Ann Arbor (MI): NSF International.

NSF International (2013b). NSF Standard 223: Conformity assessment requirements for certification bodies that certify products pursuant to NSF/ANSI 60: Drinking water treatment chemicals – Health effects. Ann Arbor (MI): NSF International.

Olivieri VP, Snead MC, Krusé CW, Kawata K (1986). Stability and effectiveness of chlorine disinfectants in water distribution systems. Environ Health Perspect. 69:15–29.

Ontario Ministry of the Environment (2008). Design guidelines for drinking-water systems. Toronto: Ontario Ministry of the Environment (http://www.ontario.ca/environment-and-energy/design-guidelines-drinking-water-systems, accessed 6 May 2014).

Pedley S, Bartram J, Rees G, Dufour A, Cotruvo J, editors (2004). Pathogenic mycobacteria in water: a guide to public health consequences, monitoring and management. London: IWA Publishing on behalf of the World Health Organization (http://www.who.int/water_sanitation_health/emerging/pathmycobact/en/, accessed 6 May 2014).

Pisigan RA, Singley JE (1987). Influence of buffer capacity, chlorine residual, and flow rate on corrosion of mild steel and copper. J Am Water Works Assoc. 79(2):62–70.

REFERENCES

Power K, Nagy I (1999). Relationship between bacterial regrowth and some physical and chemical parameters within Sydney's drinking water distribution system. Water Res. 33(3):741–50.

Prentice M (2001). Disinfection modeling study for the Owen Sound water distribution system. In: Proceedings of the AWWA Annual Conference. Denver (CO): American Water Works Association.

Reiber SH (1989). Copper plumbing surfaces: an electrochemical study. J Am Water Works Assoc. 87(7):114.

Reiber SH (1991). Galvanic stimulation of corrosion on lead–tin solder-sweated joints. J Am Water Works Assoc. 83(7):83–91.

Reiber S (1993). Chloramine effects on distribution system materials. Denver (CO): American Water Works Association Research Foundation and American Water Works Association.

Reiber SH, Dostal G (2000). Well water disinfection sparks surprises. Opflow. 26(3):1, 4–6, 14.

Revetta RP, Gomez-Alvarez V, Gerke TL, Curioso C, Santo Domingo JW, Ashbolt NJ (2013). Establishment and early succession of bacterial communities in monochloramine-treated drinking water biofilms. FEMS Microbiol Ecol. 86(3):404–14.

Rossman LA (2000). EPANET user's manual. Cincinnati (OH): United States Environmental Protection Agency, Risk Reduction Engineering Laboratory.

Rowland A, Grainger R, Smith RS, Hicks N, Hughes A (1990). Water contamination in north Cornwall: a retrospective cohort study into the acute and short-term effects of the aluminum sulphate incident in July 1988. J R Soc Health. 110:166–72.

Sarin P, Clement JA, Snoeyink VL, Kriven WM (2003). Iron release from corroded, unlined cast-iron pipe. J Am Water Works Assoc. 95(11):85–96.

Schock MR (1989). Understanding corrosion control strategies for lead. J Am Water Works Assoc. 81(7):88–100.

Schock MR, Neff CH (1988). Trace metal contamination from brass fittings. J Am Water Works Assoc. 80(11):47–56.

Schock MR, Lytle DA, Clement JA (1995). Effect of pH, DIC, orthophosphate, and sulfate on drinking water cuprosolvency. Cincinnati (OH): United States Environmental Protection Agency, Office of Research and Development (EPA/600/R-95/085).

Schock MR, Wagner I, Oliphant RJ (1996). Corrosion and solubility of lead in drinking water. In: Internal corrosion of water distribution systems. Denver (CO): American Water Works Association Research Foundation and American Water Works Association; 131–230.

Semenza JC, Roberts L, Henderson A, Bogan J, Rubin CH (1998). Water distribution system and diarrheal disease transmission: a case study in Uzbekistan. Am J Trop Med Hyg. 59(6):941–6.

Shakoor S, Beg MA, Mahmood SF, Bandea R, Sriram R, Noman F et al. (2011). Primary amebic meningoencephalitis caused by *Naegleria fowleri*, Karachi, Pakistan. Emerg Infect Dis. 17(2):258–61.

Shi B, Taylor JS (2007). Iron and copper release in drinking-water distribution systems. J Environ Health. 70(2):29–36.

Singley JE (1994). Electrochemical nature of lead contamination. J Am Water Works Assoc. 86(7):91–6.

Sorg TJ, Schock MR, Lytle DA (1999). Ion exchange softening: effects on metal concentrations. J Am Water Works Assoc. 91(8):85–97.

Standards Australia (2005). Australian Standard AS/NZS 4020: Testing of products for use in contact with drinking water. Sydney: Standards Australia.

Standards Australia/Standards New Zealand (2003). Australian and New Zealand Standard AS/NZS 3500:1:2003: Plumbing and drainage. Part 1: Water services. Sydney: Standards Australia; and Wellington: Standards New Zealand.

States S, Tomko R, Scheuring M, Casson L (2002). Enhanced coagulation and removal of *Cryptosporidium*. J Am Water Works Assoc. 94(11):67–77.

St Louis ME (1988). Water-related disease outbreaks, 1985. MMWR CDC Surveill Summ. 37(2):15–24.

Thomas PM, editor (1990). Chloramination of water supplies. Melbourne: Urban Water Research Association of Australia (Research Report No. 15).

Umweltbundesamt (2012). The 4MS Initiative: co-operation on the development of a common approach to the hygienic approval of products in contact with drinking water. Dessau-Rosslau: Umweltbundesamt.

USEPA (United States Environmental Protection Agency) (1992). Control of biofilm growth in drinking water distribution systems. Washington (DC): United States Environmental Protection Agency (EPA/625/R-92/001).

USEPA (United States Environmental Protection Agency) (1999). Alternative disinfectants and oxidants guidance manual. Washington (DC): United States Environmental Protection Agency (EPA 815-R-99-014).

USEPA (United States Environmental Protection Agency) (2002a). Potential contamination due to cross-connections and backflow and the associated health risks. Washington (DC): United States Environmental Protection Agency.

USEPA (United States Environmental Protection Agency) (2002b). Permeation and leaching. Washington (DC): United States Environmental Protection Agency.

USEPA (United States Environmental Protection Agency) (2002c). Effects of water age on distribution system water quality. Washington (DC): United States Environmental Protection Agency.

USEPA (United States Environmental Protection Agency) (2002d). Nitrification. Washington (DC): United States Environmental Protection Agency.

USEPA (United States Environmental Protection Agency) (2002e). Health risks from microbial growth and biofilms in drinking water distribution systems. Distribution System White Paper. Washington (DC): United States Environmental Protection Agency, Office of Water, Office of Ground Water and Drinking Water (http://www.epa.gov/ogwdw000/disinfection/tcr/pdfs/whitepaper_tcr_biofilms.pdf, accessed 6 May 2014).

USEPA (United States Environmental Protection Agency) (2002f). The effectiveness of disinfectant residuals in the distribution system. Distribution System White Paper. Washington (DC): United States Environmental Protection Agency, Office of Water, Office of Ground Water and Drinking Water (http://www.epa.gov/ogwdw/disinfection/tcr/pdfs/issuepaper_effectiveness.pdf, accessed 6 May 2014).

REFERENCES

USEPA (United States Environmental Protection Agency) (2006). Distribution system indicators of drinking water quality. Total Coliform Rule Issue Paper. Washington (DC): United States Environmental Protection Agency, Office of Water, Office of Ground Water and Drinking Water (http://www.epa.gov/ogwdw/disinfection/tcr/pdfs/issuepaper_tcr_indicators.pdf, accessed 6 May 2014).

USEPA (United States Environmental Protection Agency) (2007a). Elevated lead in D.C. drinking water – A study of potential causative events. Final summary report. Washington (DC): United States Environmental Protection Agency, Office of Water (EPA 815-R-07-021).

USEPA (United States Environmental Protection Agency) (2007b). Simultaneous compliance guidance manual for the long term 2 and stage 2 disinfection byproduct rules. Washington (DC): United States Environmental Protection Agency, Office of Water (EPA 815-R-07-017).

USEPA (United States Environmental Protection Agency) (2010). Control and mitigation of drinking water losses in distribution systems. Washington (DC): United States Environmental Protection Agency, Office of Water (EPA 816-D-09-001; http://water.epa.gov/type/drink/pws/smallsystems/upload/Water_Loss_Control_508_FINALDEc.pdf, accessed 6 May 2014).

Van Lieverloo JHM, Blokker EJM, Medema G, Hambsch B, Pitchers R, Stanfield G et al. (2006). MicroRisk: contamination during distribution. Performed as part of the MicroRisk project that is co-funded by the European Commission under the Fifth Framework Programme, Theme 4: "Energy, environment and sustainable development" (http://www.microrisk.com/uploads/microrisk_distribution_assessment.pdf, accessed 6 May 2014).

Van Lieverloo JHM, Blokker EJM, Medema G (2007). Quantitative microbial risk assessment of distributed drinking water using faecal indicator incidence and concentrations. J Water Health. 5(Suppl 1):131–49.

Vik E, Ryder R, Wagner I, Ferguson F (1996). Mitigation of corrosion effects. In: Internal corrosion of water distribution systems. Denver (CO): American Water Works Association Research Foundation and American Water Works Association.

Water Services Association of Australia (2002). Water Supply Code of Australia WSA 03-2002. Melbourne and Sydney: Water Services Association of Australia.

Water Services Association of Australia (2005). Dual Water Supply Systems Version 1.2. A supplement to the Water Supply Code of Australia WSA 03-2002. Melbourne and Sydney: Water Services Association of Australia.

Westrell T, Bergstedt O, Stenstrom TA, Ashbolt NJ (2003). A theoretical approach to assess microbial risks due to failures in drinking water systems. Int J Environ Health Res. 13(2):181–97.

WHO (World Health Organization) (2004). Monochloramine in drinking-water. Background document for development of WHO Guidelines for Drinking-water Quality. Geneva: World Health Organization (http://www.who.int/water_sanitation_health/dwq/chemicals/en/monochloramine.pdf?ua=1, accessed 30 June 2014).

WHO (World Health Organization) (2008). N-Nitrosodimethylamine in drinking-water. Background document for development of WHO Guidelines for Drinking-water Quality. Geneva: World Health Organization (http://www.who.int/water_sanitation_health/dwq/chemicals/ndma_2add_feb2008.pdf?ua=1, accessed 30 June 2014).

WHO (World Health Organization) (2011). Guidelines for drinking-water quality, fourth edition. Geneva: World Health Organization (http://www.who.int/water_sanitation_health/publications/2011/dwq_guidelines/en/, accessed 6 May 2014).

WHO (World Health Organization) (2012). Water safety planning for small community water supplies: step-by-step risk management guidance for drinking-water supplies in small communities. Geneva: World Health Organization (http://www.who.int/water_sanitation_health/publications/2012/water_supplies/en/, accessed 6 May 2014).

WHO (World Health Organization) (in preparation). Quantitative microbial risk assessment for water safety management: a harmonized approach to the implementation of QMRA in the water-related context. Geneva: World Health Organization.

WHO (World Health Organization), WPC (World Plumbing Council) (2006). Health aspects of plumbing. Geneva: World Health Organization (http://www.who.int/water_sanitation_health/publications/plumbinghealthasp/en/, accessed 6 May 2014).

Wilczak A, Assadi-Rad A, Lai HH, Hoover LL, Smith JF, Berger R et al. (2003). Formation of NDMA in chloraminated water. J Am Water Works Assoc. 95(9):94–106.

Wingender J, Flemming HC (2011). Biofilms in drinking water and their role as reservoir for pathogens. Int J Hyg Environ Health. 214:417–23.

Wolfe RL, Lieu NI, Izaguirre G, Means EG (1990). Ammonia-oxidizing bacteria in a chloraminated distribution system: seasonal occurrence, distribution, and disinfection resistance. Appl Environ Microbiol. 56(2):451–62.

Wong CS, Berrang P (1976). Contamination of tap water by lead pipe and solder. Bull Environ Contam Toxicol. 15(5):530–34.

Yoder J, Roberts V, Craun GF, Hill V, Hicks L, Alexander NT et al. (2008). Surveillance for waterborne-disease outbreaks associated with drinking water and water not intended for drinking – United States, 2005–2006. MMWR CDC Surveill Summ. 57(9):39–69.

CASE-STUDY ANNEXES

Case-study 1:
Application of a predictive model for water distribution system risk assessment in India

A1.1 Study area

Nagpur, the largest city in central India, is spread over 220 km^2 and has a population of about 2.5 million, as per the 2011 census. The water supply system in Nagpur is regulated by the Water Works Division of the Nagpur Municipal Corporation through a private operator, Orange City Water. Total water supply to the city is 540 million litres per day (MLD), which is distributed through 2100 km of pipeline. A water safety plan (WSP) was prepared in November 2011 and later evaluated in August 2013. An improved risk assessment for water distribution system (IRA-WDS) model was applied in the Reshimbagh and Wanjari Nagar areas of Nagpur to determine the risk probability of the distribution system. This area, bounded by 79°5″15′–79°6″30′ E and 21°7″20′–21°8″15′ N, is mainly residential and receives intermittent (1–2 hours/day) water supply. The distribution network is approximately 65 km in length, and pipe is placed 1–1.7 m below ground level. There are 961 household connections, in addition to a few connections for commercial centres and educational institutions. Sewage is carried through a 25- to 30-year-old sewer network, which is approximately 47 km in length.

In the present case-study, IRA-WDS, a geographic information system (GIS)–based software, was used for the assessment of distribution systems, the determination of risk probability, the identification of high-risk areas and the rehabilitation of distribution systems using modelling results. The following sections describe the model setup and application, utilization of modelling results in the rehabilitation of distribution systems, results and lessons learnt.

A1.2 Application of the IRA-WDS

IRA-WDS aids in evaluating the risk of deterioration of a piped water distribution network and contaminant intrusion in a water supply system in urban areas of developing countries. The software has three models – namely, the contaminant ingress model, the pipe condition assessment model and the risk assessment model – that may be used together or individually. The models are run with input of system-specific data on over 20 attributes of pipes, such as installation year, diameter, material, length, bury depth, traffic conditions and location of valves, along with sewer and drain networks. Pipe breaks and bursts are considered to be a function of installation year, traffic load and material type. GIS mapping and risk assessment lead to ranking of pipes as very high risk, high risk, medium risk or low risk. Control measures are recommended accordingly.

A1.2.1 GIS mapping
Data such as water supply network, sewer network, open drains, groundwater table, pressure in the pipes, soil characteristics and other system-specific attributes are incorporated in ArcGIS 9.3 software. Groundwater quality baseline assessment is carried out by collecting and analysing samples using *Standard Methods for the Examination of Water and Wastewater* (Rice et al., 2012). The distribution system is modelled as interconnected links and nodes in the GIS. This facilitates generation of spatial attributes, such as geographic coordinates, pipe length and area, whereas the system-specific attributes,

such as pipe material, diameter and bury depth, are generated from data records available from the Nagpur Municipal Corporation. The data are verified by field survey and interaction with field staff. Typical characteristics of the system are as follows:

- *Sewer*: the network of reinforced cement concrete and stoneware pipes is typically placed 1.35–1.85 m below the ground;
- *Soil*: mostly clay;
- *Groundwater*: water quality mostly meets Indian Drinking Water Quality Standard BIS IS 10500:2012. Groundwater table ranges from 3.1 to 10.3 m; and
- *Pressure*: Water supply pressures in different areas vary from 0.1 to 1.5 m during supply hours.

A1.2.2 Model parameter estimation

Records of operation and maintenance are used to quantify the breakage frequency and leakage in the pipe. Out of a total 1.97 MLD of water supply in the area, 1.166 MLD was getting lost prior to rehabilitation through leakage in the distribution system and unauthorized consumption. The average number of pipe breaks and bursts is 188 per year.

A1.2.3 Field survey

Hazardous events identified during field surveys of the distribution system include poor physical conditions, traffic load, open drains, crossing of sewer and water supply networks, intermittent water supply, back-siphoning during non-supply hours, illegal connections, tampering, and irregular operation and maintenance.

A1.3 Results

Fig. A1.1 presents a map showing the relative risk ranking of each pipe in the network generated by the IRA-WDS model. Risk due to contaminant ingress is determined by the presence of contaminant zones near leaky sewers and open drains and pipe condition, which in turn is determined by the pipe condition assessment model. Accordingly, the relative risk ranks indicate that 3–4% (i.e. 47) of pipes are in high to very high risk categories (risk ranks 2 and 1, respectively, in Fig. A1.1), approximately 6–7% of pipes are in the medium risk category (risk rank 3 in Fig. A1.1) and 89.7% are in the low risk category (risk rank 4 in Fig. A1.1).

Water quality analysis at the high-risk points indicated a range of faecal coliform concentrations from 18 to 126 colony-forming units per 100 mL, verifying the model results.

A1.3.1 Utilizing modelling results

A 24 × 7 water supply scheme is being implemented in Nagpur city, and the recommendations of the WSP team are being used to prioritize interventions. Modelling results of the Reshimbagh and Wanjari Nagar areas of Nagpur were discussed with the Nagpur Municipal Corporation and Orange City Water in prioritizing pipe replacement according to risk probabilities. The Nagpur Municipal Corporation and Orange City Water prioritized rehabilitation of 3% of the pipes that were at high risk to minimize the number of cases of pipe bursts, leaks, breakages and unaccounted-for water. During this period, regular monitoring and maintenance of 7% of the pipes in medium-risk condition were carried out. In the subsequent step, rehabilitation of these medium-risk pipes was also carried out. The rehabilitation programme included replacement of 1400 m of old cast iron pipes with high-density polyethylene pipes, which was completed in November 2013. Although the data on reductions in unaccounted-for water are still being monitored, it is estimated by the Nagpur Municipal Corporation and Orange City Water that the rehabilitation of pipes resulted in a 15–25% reduction in water leakages.

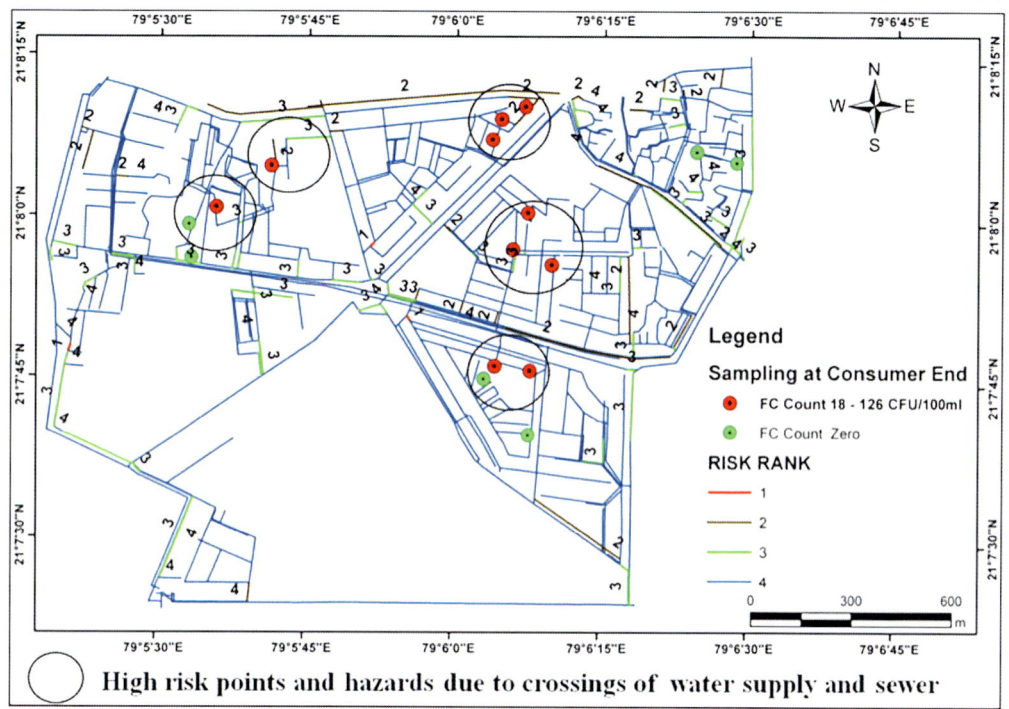

Fig. A1.1. Mapping and verification of high-risk points in pilot area. CFU: colony-forming units; FC: faecal coliforms.

Water quality analysis subsequent to rehabilitation of the distribution system indicated a faecal coliform count of zero at all eight locations where water samples were collected. These samples were collected from the same locations that reported contamination prior to rehabilitation.

A1.4 Lessons learnt

- Mathematical modelling is an important tool in identifying contaminant zones and determining risk probability in the distribution system.
- Results of mathematical modelling facilitated prioritization in rehabilitating high- and medium-risk pipes.
- Water quality analysis confirmed the integrity of the system subsequent to the rehabilitation programme.

A1.5 References

Rice EW, Baird RB, Eaton AD, Clesceri LS, editors (2012). Standard methods for the examination of water and wastewater, 22nd edition. Washington (DC): American Water Works Association/American Public Works Association/Water Environment Federation.

Linkages to WSP steps

Step 1: Assemble the WSP team

Considering Reshimbagh and Wanjari Nagar as the study areas, the WSP team included operation and maintenance heads of the Nagpur Municipal Corporation and Orange City Water from the respective areas and scientists from the National Environmental Engineering Research Institute.

Step 2: Description of the system

The water distribution network, sewer network, roads and drains were mapped and described with respect to location and pipe properties.

Step 3: Identification of hazardous events and risk assessment

Various hazardous events, such as open drains, crossing of sewer and water supply networks, intermittent water supply, back-siphoning during non-supply hours and illegal connections, were identified through field survey, water quality monitoring and discussion with Nagpur Municipal Corporation and Orange City Water staff.

Step 4: Determination and validation of control measures

Control measures included rehabilitation of the existing cast iron pipes with high-density polypropylene pipes. High-density polypropylene material is known to have excellent fluid transmission properties, which include hardness, rigidity, high tensile strength, resistance to abrasion and chemical attack, and high melting point. Additionally, the joints provided as per the new technique of electro-fusion are highly efficient.

Step 5: Develop, implement and maintain improvement plan

Ranking helped in prioritizing the very high risk/high risk pipelines to be considered for immediate replacement, recommended under the improvement plan. The improvement plan was prepared and executed during the rehabilitation programme.

Step 6: Define monitoring of control measures

Operational monitoring included proper installation of new pipelines, ensuring gradient, distance from sewer lines, drilling or excavating to sufficient depth without hampering any other service connection, testing water quality and flow after laying, appointing skilled personnel at respective locations, etc.

Step 7: Verification of implementation

Implementation was carried out under the supervision of field engineers. Progress was conveyed to the WSP team. Reduction in consumer complaints suggested satisfaction at the user end. Complaints recorded were reported to be addressed in a short time. Verification of the rehabilitation programme was confirmed through water quality checks, reduction in unaccounted-for water and improvement in other service-level benchmarks identified by the Government of India.

Step 8: Preparing management procedures

Standard operating procedures were developed and implemented. Adherence to the same for pipe installation and regular water quality analysis ensured the safety of supplied water.

Step 9: Developing supporting programmes

Supporting programmes include a refresher course on WSPs for the field supervisors.

Steps 10 and 11: Periodic revision and review

A review of the rehabilitation work was carried out by the team from the National Environmental Engineering Research Institute to update the information and include the same in the revision of the WSP document, as well as convey to other stakeholders in the Nagpur WSP as a whole. The review also enabled validation of the model in the two localities.

Case-study 2:
Distribution network management utilizing the block system to reduce non-revenue water in Phnom Penh, Cambodia

A2.1 Water supply in Phnom Penh

The water supply system in Phnom Penh, the capital of Cambodia, was constructed from 1895 to around 1966, and its capacity was 140 000 m^3/day. However, water supply facilities suffered catastrophic damage during decades of civil war. At the end of the civil war in 1991, about 85% of Phnom Penh residents were deprived of water supply services.

A massive water supply reconstruction project was started in 1992 with the assistance of Japan, France, the World Bank and the Asian Development Bank, among others. The Japanese government donated about US$ 75 million and led this project.

In March 2004, the reconstruction project was completed. Phnom Penh residents once again had a water supply and were free from the labour-intensive drawing of water.

The most important issue facing the Phnom Penh Water Supply Authority was to appropriately operate and maintain these brand new water supply facilities constructed by the project at great expense.

A2.2 Block distribution system and data monitoring system

In 2000, the Phnom Penh Water Supply Authority decided to introduce the block distribution system and data monitoring system as tangible methods for the proper operation and maintenance of the distribution facilities, as proposed by the Water and Sewer Bureau, City of Kitakyushu, Japan. To assist in the above, the Water and Sewer Bureau, City of Kitakyushu, dispatched experts and provided the necessary equipment under the Japan International Cooperation Agency project. In December 2003, the preparation of both systems was completed.

The block distribution system in Phnom Penh divides the water supply area, which consisted of 1300 km of pipelines, into 41 blocks, each with one inlet point. A flow meter and pressure meter were installed at the inlet point of each block.

The data monitoring system records the flow and pressure every minute and transmits the data to the central computer unit 3 times per day, utilizing telemeter and telephone lines.

These data are summarized in a daily report, including the daily supply amount, the maximum and nighttime minimum hourly supply amounts and the time at which these maximum and minimum supply amounts occurred. Also, graphs of flow and pressure trends in the 41 blocks are created every day.

A2.3 The Project on Capacity Building for Urban Water Supply System in Cambodia (Phase 1)

The Japan International Cooperation Agency's technical cooperation project was implemented for 3 years starting in October 2003. This project was conducted with the Phnom Penh Water Supply Authority. Its main purposes were (1) non-revenue water reduction, (2) appropriate water treatment operation, (3) improvement of water quality and (4) appropriate operation and maintenance of electrical and mechanical facilities. The challenges of the water distribution sector are to properly operate and maintain the newly developed block distribution system and data monitoring system and to reduce the non-revenue water utilizing these systems. The analytical skills and knowledge needed to collect the data from the 41 blocks by the data monitoring system and the effective water leakage detection skills were transferred through on-the-job training and the preparation of standard operating procedures together with the Phnom Penh Water Supply Authority under the project.

At the end of the project, in 2006, the non-revenue water ratio had been improved to 7.3% (Fig. A2.1), which is almost the same level as for the water supply in Japan.

Also, awareness of water quality increased. The Phnom Penh Water Supply Authority supplies drinkable water to the city at all hours. Water quality analysis, which includes temperature, pH, turbidity, colour, conductivity, residual chlorine and coliforms, is conducted at the 27 taps in the water supply area once a week.

The non-revenue water reduction secures financial sustainability by keeping the water tariff unchanged since 2001, making the water affordable to millions of the most vulnerable people in Phnom Penh.

This reconstruction and technical cooperation project in Phnom Penh, Cambodia, was very successful, based on its results (Table A2.1).

Fig. A2.1. Non-revenue water reduction in Phnom Penh since 1993

Table A2.1. Rehabilitation and expansion of water supply capacity of the Phnom Penh Water Supply Authority

Parameter	1993	2005
Supply duration	10 h/d	24 h/d
Pressure	2 m	25 m
Coverage area	25%	95%
Number of connections	26 881	138 000
Network length	280 km (old)	1 230 km (new)
Production	65 000 m^3/d	235 000 m^3/d
Non-revenue water	72%	11%
Metering	12%	100%
Collection	48%	100%

Source: Visoth Chea, Appropriate Operation and Maintenance of Water Supply and Sanitation Facilities – Phnom Penh's Experience, 4th World Water Forum, Mexico City, March 2006

Linkages to WSP steps

Step 6: Define monitoring of control measures

Operational monitoring, including water quality analysis, was conducted at 27 taps in the water supply area once a week. Parameters monitored included temperature, pH, turbidity, colour, conductivity, residual chlorine and coliforms.

Step 8: Preparing management procedures

Standard operating procedures were developed and implemented.

Step 9: Developing supporting programmes

Supporting programmes included training on the understanding and use of the newly implemented block distribution system and data monitoring system.

Case-study 3:
Water safety plans and distribution systems – the case of Spanish Town, Jamaica

A3.1 Introduction

The first water safety plan (WSP) undertaken in the Western Hemisphere was implemented in Spanish Town, Jamaica. The WSP in Spanish Town was initiated by the Environmental Health Unit within the Jamaican Ministry of Health and the National Water Commission of Jamaica (the service provider), with technical assistance from the United States Centers for Disease Control and Prevention, the Pan American Health Organization, the United States Environmental Protection Agency and the Caribbean Environmental Health Institute (now part of the Caribbean Regional Public Health Agency) (Environmental & Engineering Managers Ltd, 2007). In order to guide the WSP, a WSP Task Force was created, including various Jamaican organizations involved in issues related to water supply. The Task Force was chaired by the Water Resources Authority of Jamaica and the Ministry of Health/Environmental Health Unit and also included the National Water Commission, National Irrigation Commission, National Environment & Planning Agency, National Public Health Laboratory, Pesticides Control Authority, the Parish Council and Health Department from the Parish of St Catherine in which Spanish Town is located, and a private consultant to assist with coordinating the task force.

A3.2 Identifying and characterizing risks to drinking-water safety

One of the first steps undertaken by the WSP Task Force was to characterize the drinking-water system in Spanish Town. In 2006, when the WSP was initiated, the Spanish Town drinking-water system served an estimated 140 000 people with 10 groundwater wells and surface water supplied by a canal carrying water diverted from the Rio Cobre. Groundwater was chlorinated at the wellheads, and surface water was treated in a conventional drinking-water treatment plant before entering the distribution system. A 2002 inventory identified a total of 312.6 km of pipe in the distribution system. Losses in the distribution system occurred from unauthorized connections, non-functioning meters or unmetered supplies (e.g. public standpipes) and leakage, all contributing to unaccounted-for water.

The WSP Task Force created a methodology for identifying and ranking risks to drinking-water safety based on a scale with four levels of risk: very high, high, moderate and low. In addition, risks were classified as to where they occurred: catchment, treatment, storage and distribution. As shown in Table A3.1, the Task Force identified 44 distinct risks, with 25% (11/44) of those risks being in the distribution system.

Table A3.1. Number and location of risks identified in the Spanish Town water system

	Catchment	Treatment	Storage	Distribution
No. of risks identified	15	16	2	11
No. of very high risks identified	2	1	–	–
No. of high risks identified	1	3	1	7

Table A3.1 also shows how the very high and high risks identified by the Task Force were distributed throughout the water system. No very high risks were identified in the distribution system; however, a large proportion of the high risks identified were in the distribution system. The following high risks were identified in the distribution system (Environmental & Engineering Managers Ltd, 2007):

- inability to supply minimum of 6 hours of water if treatment plant shuts down;
- unauthorized connections in distribution system;
- high percentage of unaccounted-for water due to lack of metering/malfunctioning meters;
- leaks in mains and distribution system;
- demand exceeding supply;
- low pressure if all sources not in service; and
- backflow prevention in households absent or ineffective.

A3.3 Remedial actions

A number of remedial actions were undertaken in response to these findings from the WSP in Spanish Town. Some of these were part of larger improvement efforts that were already under way, whereas others were directly as a result of the WSP. Risks throughout the drinking-water system were addressed, but only those related to the distribution system are discussed here (more comprehensive information is available from the Spanish Town WSP document – see Environmental & Engineering Managers Ltd, 2007). Remedial actions related to the distribution system included upgrading the capacity of the surface water treatment plant (to help ensure more consistent supply), water main replacement, installation of zone meters in the distribution system and refurbishment of water storage tanks. Other actions related to unauthorized connections and improving cross-connection control were also undertaken. Because it was not always politically acceptable to cut service for unauthorized connections, water service was maintained at lower pressures in some areas, so as to provide service but not to waste too much water. In order to improve cross-connection control, new procedures were implemented to enforce backflow control on new construction and work with private sector suppliers to prevent backflow from household water storage tanks.

A3.4 Lessons learnt

Results from the WSP in Spanish Town, as well as other WSPs in the Latin America and the Caribbean region, showed several trends related to distribution system issues. In the context of these WSPs, multiple risks in the distribution system were identified consistently. Those risks were not always considered the very highest risks in those systems, but they did constitute high risk factors. This reflects the multiple water quality challenges in these drinking-water systems, where source contamination and inadequate treatment are often issues.

Of the risks identified in distribution systems, intermittent service and illegal connections were consistently identified as risks to drinking-water safety. Some of these distribution risks are linked to other components within drinking-water systems. For example, intermittent service may be related to insufficient quantities of water from sources in addition to losses of water in distribution. Irregular monitoring of water quality and inconsistent use of monitoring data were also identified as distribution system risks.

Results from this and other WSPs have shown that the WSP process is an effective means for identifying risks in distribution systems, as well as mitigation strategies to deal with those risks. Results also indicated that the added value of a WSP was in helping to make management of water supply systems more systematic and comprehensive and focus scarce resources on the highest-risk issues.

Because WSPs are a stakeholder-based process designed to gather input from various partners beyond solely water utilities, they are effective at identifying a broad spectrum of risks to water safety. In addition, because they draw together various partners involved in protecting water safety, WSPs also improve communication and collaboration between key stakeholder groups. This, in turn, improves stakeholders' understanding of water systems and their role in protecting water quality, including in distribution systems.

A3.5 References

Environmental & Engineering Managers Ltd (2007). Water safety plan, Spanish Town water supply, St Catherine, Jamaica, October 2007. Kingston, Jamaica.

Linkages to WSP steps

Step 1: Assemble the WSP team

At the beginning of the project, a WSP Task Force was created that included various Jamaican organizations involved in issues related to water supply. These included the Environmental Health Unit within the Jamaican Ministry of Health, the National Water Commission, the Water Resources Authority, National Irrigation Commission, National Environment & Planning Agency, National Public Health Laboratory, Pesticides Control Authority, the Parish Council and Health Department from the Parish of St Catherine in which Spanish Town is located, and a private consultant.

Step 2: Description of the system

One of the first steps undertaken by the WSP Task Force was to develop a comprehensive description of the drinking-water supply system in Spanish Town. The system description was broken down by components of the water system — namely, catchment, treatment, storage and distribution.

Step 3: Identification of hazardous events and risk assessment

The WSP Task Force created a methodology for identifying and ranking risks to drinking-water safety based on a scale with four levels of risk: very high, high, moderate and low. In addition, risks were classified as to where they occurred: catchment, treatment, storage and distribution.

Step 4: Determination and validation of control measures

The WSP Task Force identified both existing and potential control measures to deal with the risks that had been identified and prioritized. Risks were dealt with in priority order, with the very high and high risks being addressed first.

Step 5: Develop, implement and maintain improvement plan

Improvement plans were developed to implement remedial actions related to the distribution system (high-priority risks in other components of the water system were also addressed). Actions included upgrading the surface water treatment plant, replacement of water mains, installation of master meters, rehabilitation of water storage tanks, and control of unauthorized connections and cross-connections.

Step 6: Define monitoring of control measures

The WSP identified control measures for each risk and the organization or individuals responsible for monitoring control measures, as well as corrective actions to be taken if control measures were found not to be effective.

Step 7: Verification of implementation

Verification procedures for distribution system issues identified by the WSP were also identified. Many of these relied largely on existing monitoring procedures, such as internal standard operating procedures (SOPs) of the National Water Commission and water quality inspections by the Ministry of Health. Nonetheless, the WSP helped to provide a structure for these existing efforts and also identified time frames for monitoring of control measures.

Step 8: Preparing management procedures

Many SOPs had already been developed by the National Water Commission, but the WSP process documented and formalized the SOPs. These documented SOPs will help to ensure that all National Water Commission staff are informed of proper procedures and follow them.

Step 9: Developing supporting programmes

To support future development of additional WSPs in Jamaica, the National Water Commission implemented training on WSPs for all National Water Commission offices throughout Jamaica. The WSP Task Force also recommended that a specific organizational unit within the Ministry of Health be designated to monitor implementation of the Spanish Town WSP.

Steps 10 and 11: Periodic revision and review

The National Water Commission planned to review and update the WSP annually and implement external audits of WSP implementation in the future.

Case-study 4:
Drinking-water contamination incident in Johannesburg, South Africa

A4.1 Introduction

Johannesburg Water is an entity of the City of Johannesburg, South Africa, responsible for water and sanitation services within the city precinct. The utility provides services to 4.4 million consumers through an extensive distribution network of over 22 000 km of reticulation.

A4.2 Sewer line refurbishment leads to water contamination

As part of Johannesburg Water's sewer refurbishment programme, a contractor was employed to replace and upgrade 2.6 km of existing sewer lines of varying pipe diameters and material within the vicinity of the Diepsloot Township, located to the north-west of the City of Johannesburg.

In April 2012, the appointed contractor was excavating a section of the sewer network that was approximately 4 m deep and replacing existing pipelines and manholes. This particular section of the sewer network was adjacent and parallel to a potable water pipeline feeding the Diepsloot area. When one of the sewer manholes was removed as part of the construction, an adjacent air valve of the water main moved, resulting in a break in the water main. Due to the break in the water main and the ongoing excavations on the sewer main, there was an immediate and apparent intermixing of water and sewage, leading to water contamination. Piped water supply was therefore interrupted to the whole of the Diepsloot area for 10 days, and an alternative supply was provided in water tankers for the duration of the incident. Fortunately, there were no known cases of customers exposed to the contaminated water, as the water pipe break meant a disruption of service, and service was resumed only when it was ascertained that the water within the network was suitable for human consumption.

Although there were other interventions and lessons learnt, this case-study focuses only on the infrastructure activities that led to the drinking-water contamination.

A4.3 Root causes

A4.3.1 Irregular network layout and fixtures

The horizontal distance between the water and sewer pipelines in this scenario was about 1.5 m, which is quite close, whereas there was available space within the area that could have accommodated an acceptable distance between the two pipelines.

Furthermore, the thrust block of the air valve (water pipeline) was built against the sewer manhole, as shown in Fig. A4.1, which shows the temporary reconstruction. This not only made the sewer and water networks interdependent, but turned the sewer manhole into a structural support that was to bear any thrusts and surges from the water pipeline. Both these factors do not conform to standard practice.

Another difficulty experienced when staff were trying to isolate the water network was the very low number of isolation valves that could be found and operated and the distance between the valves.

Fig. A4.1. Water pipeline thrust block supported by a sewer manhole

A4.3.2 Inaccurate as-built drawings

Assuming the foregoing were exceptions that could have been necessary, the as-built drawings would clearly indicate each of these fixtures with the necessary warnings. However, only the proximity of the pipelines to each other was indicated on the drawings, without any information on how the thrust block was anchored. Although the contractor could possibly have exposed the manholes prior to demolishing them, there was no perceived need to do so – hence the adopted plan of action. Had the thrust blocks been indicated on drawings, care would have been taken to ensure minimum disruption to services.

A4.3.3 Network operation and maintenance

A few isolating valves could be found and operated. Lack of valve maintenance also played a critical role, as some valves were buried underground and could not be found during the drinking-water contamination incident.

A4.3.4 Contractor capacity

While the contractor was not necessarily the root cause of the incident, his response to the incident was less than adequate, mostly due to lack of capacity in both experience and resources. This negatively impacted the initial handling of the incident and lengthened the time between the incident and restoration of services.

A4.5 Incident management

A4.5.1 Establishing a crisis management team

The incident was of a magnitude requiring immediate attention and was accorded the highest level of emergency due to the extent of the problem. The crisis management team was made up of all critical functions, as documented on the disaster management plan. The team was led by executive management of the company, and it was composed of operations, scientific services, capital investment, communications and stakeholder relations, among others. The team basically managed all activities aimed at restoring services and water quality to the Diepsloot area. These included constant communicating with the public through various media outlets on progress in restoring the services. A decision was made to supply water through alternative means (tankers), as there was no secondary pipe feed to the area.

A4.5.2 Repairs to pipeline and restoration of service

The water pipeline was repaired by the in-house teams in tandem with the sewer network component, which was done by the contractor. The repairs done were temporary in nature, and the sewer network is planned for relocation as part of the annual sewer refurbishment programme. An extended monitoring programme to monitor the quality of the water was instituted over the period, together with secondary dosing of chlorine. The main challenge with bacteriological monitoring was the 24-hour waiting period for the reporting of results before any decisions could be made. The service was finally restored after 8 days when zero *E. coli* was consistently recorded.

A4.6 Lessons learnt

The following were some of the key lessons from the incident that Johannesburg Water has incorporated into its daily operations to minimize the recurrence of a similar incident:

- Correct designs, construction and accurate as-built information are invaluable for the correct operation and maintenance of the networks and for minimizing unintentional cross-contamination of drinking-water with sewage.
- Network operation and repairs have the greatest impact on creating and spreading contamination and can be better managed to minimize contamination. This includes the proactive maintenance of valves and other network installations.
- Extensive water networks need to have facilities that will facilitate secondary disinfection of the network.
- Rapid bacteriological testing and analyses that allow for reporting of results within a few hours could greatly assist in decision-making, and research efforts are under way.
- Partnerships with other departments are always critical in accelerating service delivery and ensuring consistent service delivery.

Linkages to WSP steps

Step 1: Assemble the WSP team

The WSP team included operation and maintenance heads and scientists in Johannesburg Water utility.

Step 2: Description of the system

The water supply system and sewer system of Johannesburg Water in the Diepsloot Township were described.

Step 3: Identification of hazardous events and risk assessment

Sewer network repair in the vicinity of the potable water pipeline with existing cracks resulted in the intermixing of water and sewage in the potable water pipeline.

Step 4: Determination and validation of control measures

Alternative water supply was provided to customers via water tankers for the duration of the incident.

Step 5: Develop, implement and maintain improvement plan; Step 6: Define monitoring of control measures

An extended chemical and bacteriological monitoring programme was activated to monitor water quality, and results were communicated to the crisis management team.

Step 7: Verification of implementation

The incident was managed by the crisis management team, which was made up of all critical functions, including operations, scientific services, capital investment, communications and stakeholder relations, among others. Implementation was carried out under the supervision of field engineers, with repairs of the water pipeline and relocation of the sewer network as part of the annual sewer refurbishment programme. Service was restored to customers only when zero *E. coli* was consistently recorded.

Step 8: Preparing management procedures

Correct design, construction and standard operating procedures were developed and implemented.

Case-study 5: Incident management in the water distribution system in Johannesburg, South Africa

A5.1 Introduction

A number of failures in the distribution system – namely, loss of adequate disinfectant residual, low water pressure, intermittent service and ageing of infrastructure – can result in the declining quality of the water supply (WHO & UNICEF, 2000). If poor sanitary conditions exist, pathogen intrusion may occur, leading to outbreaks of waterborne and water-related diseases. It is important to note that it is often not a single defect, but the combination of a number of failures in the distribution system, that results in poor water quality. Findings from a recent study (Mackintosh & Jack, 2008) involving selected local and district municipalities in South Africa (Western Cape, Free State and Eastern Cape) revealed that distribution system–related outbreaks occurred as a result of a number of factors, including improper installation practices; ageing infrastructure, resulting in pipe bursts and leaks, thus leading to pollutant ingress; inadequate flushing; faulty storage; inaccurate network/pipeline drawings, making it difficult for operators to verify locations and the proximity of buried water and sewer lines to each other; and construction and maintenance work in the distribution system, resulting in backflows. Among the reported cases, high coliform and *E. coli* counts were a common cause of water quality failures in the distribution systems in South Africa. In most cases, such incidents were due to problems associated with a drop in both flow rate and pressure, operational maintenance, resulting in reversed flow, and also pathogen intrusion. This resulted in water stagnation and, thus, depletion of residual disinfectant. In the absence of a residual disinfectant, microorganisms proliferate.

In recognition of these challenges, a Drinking Water Quality Framework for South Africa was developed to enable effective management of drinking-water quality to protect public health (DWA, 2005). The framework is based on an integrated system of approaches and procedures that address the key factors that govern drinking-water quality and safety in South Africa. To respond to acute drinking-water quality failures, a Drinking Water Failure Emergency Response model, which comprises three alert levels, has to be followed:

- *Alert Level I*: Routine problems, including minor disruptions to the water system and single-sample non-compliance (internal Water Services Authority response only);
- *Alert Level II*: Minor emergencies requiring additional sampling, process optimization and reporting/communication of the problem (internal Water Services Authority response only);
- *Alert Level III*: Major emergencies requiring significant interventions to minimize public health risk (engagement of an active Emergency Management Team).

There are a number of mechanisms by which distribution system water quality incidents are identified. These are outlined in the Incident Response Management protocol, which sets out the alert levels based on the magnitude and extent of the failure, response times, required actions, roles and responsibilities and communication vehicles, as well as the risks to public health posed by the failures. Operational alerts are triggered when the total coliforms/100 mL or *E. coli* counts exceed those stipulated in the South African National Standard (SANS 241) for drinking-water. In other cases, alerts arise from water quality queries received from consumers. Notifications of water quality incidents can also come from staff during construction, vandalism or damage to a water pipe or a sewer by an excavator, resulting in the drinking-water supply being contaminated with sewage or environmental contaminants (Phahlamohlaka & Rimmer, 2010). After an incident, it is required that appropriate documentation be produced and the incident reported accordingly. Particularly, notifications regarding the cause of the incident and the actions taken to minimize future occurrences, as well as communicating the end of an incident or emergency, are necessary activities for restoring consumer confidence and water supplier credibility after an incident and/or emergency situation.

A5.2 Incident management – Johannesburg Water case-study

Johannesburg Water is the water services provider for the City of Johannesburg and its sole shareholder, mandated the responsibility of providing water and sanitation to about 3.8 million residents of the City of Johannesburg. Drinking-water is distributed through a distribution water network of about 11 000 km of pipes and over 100 reservoirs and water towers, and wastewater is collected through another 11 000 km wastewater network. In Johannesburg Water, incident management response protocols and standard operating procedures (SOPs) are in place to ensure rapid reaction and appropriate response to incidents that may affect the quality of a drinking-water supply and pose a potential risk to human health. The precise actions to be taken to identify the cause of the problem, corrective actions to be implemented and monitoring schedules to be followed until the microbiological quality of the water again complies with the standard are stipulated in these SOPs. Johannesburg Water maintains an intensive water quality monitoring programme to ensure that the water supplied to its consumers is of acceptable quality. Johannesburg Water laboratories are accredited according to ISO/IEC 17025:2005. Internal and external quality assurance is carried out for the control and for monitoring the validity of tests and calibration, including Good Laboratory Practice. The following are examples of distribution system water quality failure case-studies at Johannesburg Water and mechanisms that were used to identify and manage these incidents.

A5.2.1 *Possible contamination through a pipe burst*

During maintenance of the sewer water mains by a contractor, a drinking-water pipe was accidentally broken, resulting in contaminants entering the drinking-water supply system. Johannesburg Water was immediately informed, and a state of emergency was declared. The water supply to the affected areas was cut off, and the hydrant was flushed on 3 consecutive days. During these 3 days of flushing, samples were taken at random locations of the affected areas to determine the extent of the contamination in the water supply network. In addition, samples were also taken at the surrounding premises after every flush. After several flushes, the water was safe for drinking, as it complied with the SANS 241 drinking-water standard.

A5.2.2 *Water supply contamination due to pipe leakages*

During routine operational monitoring, high total coliforms and *E. coli* were detected in samples collected from communal taps supplying water to one of the informal settlements in the City of Johannesburg. A recommendation that came from an initial investigation was the replacement of all the taps supplying water to the informal settlement. However, even after tap replacement, coliforms were still detected in the water samples collected from the new communal taps. As a result, a second detailed investigation was launched, and random samples were taken around the area, including the reservoir that feeds the communal taps. No coliforms were detected in the main reservoir, but leaks found on the mains supplying the informal settlement were the cause of the contamination observed. This incident was resolved by fixing the leaks and installing a new sampling point on the pipeline feeding the informal settlement.

A5.2.3 Consumer complaints

These incidents are identified following water quality complaints lodged by consumers. At Johannesburg Water, consumer complaints are logged at the city's call centre and are then escalated directly to the relevant department for a response. Typical queries received include suspected illness from the water, presence of larvae or worms in the water, and foul taste or odour in the water. Among these, the highest number of complaints received has to do with the presence of larvae/worms in the water, and these are received mostly during the rainy season. Upon receipt of the complaint, samples are collected from both the property concerned and neighbouring properties, and analyses are done. Results obtained assist in identifying the cause of the problem, implementation of corrective action and intensifying monitoring until the water complies with the standard (SANS 241). The call remains open until the consumer is satisfied with the outcome.

A5.3 References

DWA (Department of Water Affairs) (2005). A Drinking Water Quality Framework for South Africa. Johannesburg: Department of Water Affairs and Forestry (http://www.dwaf.gov.za/Documents/Other/DWQM/DWQMFrameworkDec05.pdf, accessed 9 December 2013).

Mackintosh G, Jack U (2008). Assessment of the occurrence and key causes of drinking-water quality failures within non-metropolitan water supply systems in South Africa and guidelines for the practical management thereof. Report to the Water Research Commission (WRC Report No. TT 373/08).

Phahlamohlaka V, Rimmer R (2010). Drinking water microbiological quality failure response management. National Laboratory of South Africa T&M Proceedings (http://www.nla.org.za/conferences/proceedings_archive/2010/02_Monday%2008%20Nov%202010/Papers/M206%20Violet%20Phahlamohlaka.pdf, accessed 7 May 2014).

WHO (World Health Organization), UNICEF (United Nations Children's Fund) (2000). Global water supply and sanitation assessment 2000 report. WHO/UNICEF Joint Monitoring Programme for Water Supply and Sanitation (http://www.who.int/water_sanitation_health/monitoring/globalassess/en/, accessed 20 May 2014).

Case-study 6:
Black water complaints from western parts of Singapore in the 1980s

In 1982–1983, there were numerous complaints by customers from the western parts of Singapore (Choa Chu Kang water supply zone) on black sediments found in their water.

The discoloured water was supplied from Choa Chu Kang Waterworks. A series of water samplings was carried out at customers' taps, distribution pipelines, service reservoirs, Choa Chu Kang Waterworks and reservoir raw water. Analyses of the samples indicated that these sediments were made up of:
- black precipitates formed from oxidation of dissolved manganese with ozone and chlorine;
- brown iron rust sediments from badly corroding pipelines; and
- organic matter from biofilm deposits found on the walls of pipelines.

Inspections found that these sediments were present in the distribution mains as well as the service reservoirs. Meanwhile, the quality of the treated water from Choa Chu Kang Waterworks was tested and found to be within the acceptability threshold of 0.1 mg/L for manganese under the *Guidelines for Drinking-water Quality* set by the World Health Organization (WHO).

A more in-depth investigation was conducted, and it was found that a combination of acidic raw water in Tengeh reservoir (Choa Chu Kang Waterworks' source water), together with changes in the network operations, led to discoloured water at customers' taps.

Choa Chu Kang Waterworks received raw water from three reservoirs – namely, Tengeh, Kranji and Pandan reservoirs, which were developed from previous tidal estuaries. Estuarial reservoir waters generally had higher organic and mineral content compared with waters from inland reservoirs or rivers. After the initial stage of site investigations, the source of the discoloured water was concluded to be Tengeh reservoir. The water in Tengeh reservoir appeared clearer than other raw water sources, and pumps were showing signs of corrosion. Investigations found that large quantities of peaty organic soils (2 million cubic metres) had been dredged from the bottom of the reservoir and used as reclaimed materials for the northern and southern banks of the reservoir. Exposure to air caused the sulfur contained in these materials to turn into soluble sulfates through a bacteriological process. The sulfates were then brought into the reservoir by the natural inflow of the water from the tributaries of the banks, resulting in low-pH (pH 3.5–4.1) water at Tengeh reservoir. Results of sampling of surrounding soil showed that the bank was a highly acidic area that caused the raw water in Tengeh reservoir to be acidic and to contain high chloride, sulfates, aluminium, iron and manganese. A substantial amount of manganese (≥0.1 mg/L) was also brought into the impoundment on rainy days. The acidity of the water in Tengeh caused the manganese to remain in dissolved form until it was oxidized by ozonation and became the black sediments.

At the same time, alterations were made to the water supply operations. Johor Waterworks was used to supply the shortfall in treated water for the Choa Chu Kang water supply zone whenever the supply from Choa Chu Kang Waterworks was reduced. A key difference was that treated water from Johor

Waterworks was dosed with chloramine to maintain a chlorine residual (because of the longer retention time due to the longer distance to the distribution network), whereas treated water from Choa Chu Kang Waterworks was dosed with free chlorine to maintain a chlorine residual in the network. Controlled experiments were conducted by mixing Johor Waterworks water with Choa Chu Kang Waterworks water. The results showed that concentrations of manganese were consistently higher whenever free chlorine was present in the treated water. When treated water sources from Johor Waterworks and Choa Chu Kang Waterworks were mixed in the service reservoir and network, dissolved manganese was oxidized by the free chlorine present in the network to form a black precipitate. Hence, the sudden change in water source from Choa Chu Kang Waterworks to Johor Waterworks caused biofilms adhering to the walls of pipelines with manganese dioxide and other deposits to be sloughed off. This resulted in the release of black sediments into the water network and out at the customers' taps. The combination of the above factors resulted in the detection of dirty water in the customers' supply.

Remedial actions, both short and long term, were taken to overcome the problem. To minimize the precipitation of manganese and organic matter in the treated water, temporary chemical storage tanks and dosing units were installed at Tengeh pumping station. Potassium permanganate dosing aided in oxidizing and precipitating substances for ease of removal in the sedimentation process, whereas sodium hydroxide increased the pH of raw water for optimum oxidation and flocculation processes. The cleaning programmes for clear water tanks in Choa Chu Kang Waterworks as well as in service reservoirs downstream were stepped up to keep the system clean. The network flushing programmes were also stepped up, especially the low-flow sub-network in the water supply zone. In addition, turfing for Tengeh North and South banks was carried out. The action controlled the acidic runoff, which was primarily caused by the acidic soil found at the barren lands upstream of Tengeh reservoir, and helped increase the pH of Tengeh reservoir to an acceptable range of 6.5–7, where the dissolution of manganese was contained at the source. The implementation of the long-term solutions successfully solved the "black water" problem in the Choa Chu Kang water supply zone by keeping the concentration of dissolved manganese in treated water below 0.02 mg/L. Short-term measures were no longer required after the turfing was completed.

Although the problem of black sediments occurred at the end of the water supply chain (delivery to customers), thorough investigations and studies were carried out at all process units from the raw water sources (reservoirs) to the distribution system to identify the cause of the problem. To meet the commitments to customers, short-term solutions to remove the black sediments through additional treatment processes were implemented. The permanent solution was to prevent the problem at the source.

Linkages to WSP steps

Step 1: Assemble the WSP team

The WSP team included operation and maintenance heads of the catchment and waterways department, water supply plant department, water supply network department, and scientists from the water quality office within Singapore Water Authority.

Step 2: Description of the system

The water supply chain from raw water reservoirs to water treatment plants to the water distribution network was described with respect to location.

Step 3: Identification of hazardous events and risk assessment

A combination of acidic raw water and changes in the network operations resulted in discoloured water at customers' taps.

Step 4: Determination and validation of control measures; Step 5: Develop, implement and maintain improvement plan; Step 6: Define monitoring of control measures

Short-term control measures included installation of temporary chemical storage tanks and dosing units at pumping stations, stepping up cleaning programmes for clear water tanks in Choa Chu Kang Waterworks and treated water service reservoirs downstream, and stepping up the network flushing programme. Turfing to control the acidic runoff was done as a long-term control measure to remove the problem at the source.

Step 7: Verification of implementation

Implementation was carried out under the supervision of field engineers. Progress was conveyed to the WSP team. Reduction in consumer complaints suggested satisfaction at the user end. Complaints recorded were reported to be addressed in a short time.

Step 8: Preparing management procedures

Standard operating procedures were developed and implemented.

Case-study 7:
Implementation of massive replacement programmes for unlined galvanized iron connections and unlined cast iron mains in Singapore in response to poor water quality after commissioning of the Kranji/Pandan scheme (1983–1993)

In the early 1980s, there was much feedback from customers on poor water quality (largely due to rusty, dirty, coloured or smelly water) and poor pressure (about 1300 and 3500 cases per year, respectively). Most of the feedback was from the western part of the island and resulted from water coming from the Kranji, Pandan and later from the Western Catchments reservoirs as well. Compared with the reservoir schemes from the central protected catchment area, the raw water from these estuarine reservoir schemes at coastal swamps and river estuaries inherently has higher chloride, dissolved oxygen and sulfate contents. As a result, the treated water from these estuarine reservoir schemes also contained higher chloride, dissolved oxygen, sulfate and total solids contents, which accelerated the corrosion of unlined galvanized iron connection pipes and unlined cast iron mains and resulted in poor quality water with undesirable taste and colour due to the presence of rust sediments.

An accelerated water main flushing programme was carried out to mitigate the accumulation of sediments in the distribution mains. However, this measure was not sustainable for reason of water wastage, resulting in higher unaccounted-for water. Flushing of the mains at such scale was also labour intensive and too widespread for the maintenance teams to manage. As a result, the Public Utilities Board of Singapore embarked on a comprehensive water main and connection pipe replacement programme. This included the replacement of some 182 km of unlined cast iron mains and around 76 000 unlined galvanized iron connecting pipes around the island with cement-lined ductile iron water mains and copper and stainless steel connecting pipes, respectively. The programme was carried out in three phases covering three regions – namely, the western, eastern and central areas of the island. In addition, customers (including managing agencies of buildings) were advised to replace their unlined galvanized iron service pipes within their premises with corrosion-resistant pipes to tie in with the Public Utilities Board's replacement work.

The various measures and replacement programmes implemented were found to be effective. The programmes not only significantly brought down the number of complaints in terms of poor water quality and pressure, but also helped to improve the standard of customer service provided by the Public Utilities Board, in reducing both the number of leakages and the interruption thus caused to the customers. Managing customers' expectations was one of the prime considerations under the 10-year intensive replacement programmes. Two Replacement & Diversion units (East and West) were formally set up under the then Maintenance & Repair Branch to execute the challenging tasks. Standard operating procedures were formulated especially to deal with the replacement, as it concerned work within customer premises and water supply interruptions, which could potentially be very sensitive. Public notices were served to customers affected by these mass replacement works. The majority of these procedures and forms continue to remain in use up to the present day.

Linkages to WSP steps

Step 1: Assemble the WSP team

The WSP team included operation and maintenance heads of the catchment and waterways department, water supply plant department, water supply network department and scientists from the water quality office within the Singapore Water Authority.

Step 2: Description of the system

The water supply chain from raw water reservoirs to water treatment plants to the water distribution network was described with respect to location and pipe properties.

Step 3: Identification of hazardous events and risk assessment

Raw water from the estuarine reservoir schemes contained higher chloride, dissolved oxygen and sulfate contents. As a result, the treated water also contained higher chloride, dissolved oxygen, sulfate and total solids contents, which accelerated the corrosion of unlined galvanized iron connection pipes and unlined cast iron mains. The rust sediments in the water resulted in offensive taste and colour.

Step 4: Determination and validation of control measures

The network flushing programme was stepped up and accelerated as a short-term control measure. As a long-term solution, the Public Utilities Board embarked on a comprehensive water main and connection pipe replacement programme to replace unlined cast iron mains and unlined galvanized iron connecting pipes with cement-lined ductile iron island-wide. Building owners were also advised to replace their unlined galvanized iron pipes with corrosion-resistant pipes.

Step 5: Develop, implement and maintain improvement plan

The island-wide water main replacement programme was carried out in three phases. Ranking helped in prioritizing the pipelines for replacement. The higher-risk pipelines were immediately replaced.

Step 6: Define monitoring of control measures

The number of customers' complaints in terms of poor water quality and pressure and the number of leakages and service interruptions to customers were monitored.

Step 7: Verification of implementation

Implementation was carried out under the supervision of field engineers. Progress was conveyed to the WSP team. A significant reduction in consumer complaints suggested satisfaction at the user end.

Step 8: Preparing management procedures

Standard operating procedures for water main replacement were developed and implemented, as the replacement concerned work within customer premises and water supply interruptions.

Step 9: Developing supporting programmes

Training programmes were developed to train the field operators on the standard operating procedures.

Case-study 8: Contamination of water supply incident in Bukit Timah Plaza/Sherwood Tower Condominium in Singapore in 2000

On 18 August 2000, a rare contamination incident occurred at a shopping cum residential complex – Bukit Timah Plaza. This complex comprises the Sherwood Tower Condominium on the upper floors and the shopping complex and eating outlets on the lower floors. The then Ministry of the Environment of Singapore received a number of complaints from a local clinic operating in the affected building, reporting an abnormal increase in the number of patients suffering from abdominal pain, diarrhoea and vomiting. An alert doctor observed that all the patients with such symptoms were either residents of the condominium or office workers from the building. The contamination was confined to this complex and did not affect Singapore's public water supply system. This incident, with about 120 people falling sick, naturally attracted strong media interest.

Site investigations carried out by the Public Utilities Board showed that the water in the low-level tank located in the building's basement was contaminated. Contaminated water was subsequently supplied to the customers through the low-level and high-level tanks. The causes of the contamination were corrosion from the sewer pipes (i.e. pipes conveying wastewater) located above the basement's low-level water tank, a corroded metal trough directly beneath the sewer pipes and a poorly maintained low-level water tank in the basement. Wastewater seeped from the leaking sewer pipe through the corroded metal trough and found its way into the water in the low-level tank through cracks and gaps on the roof of the tank. Had the tank been properly maintained, contamination would not have occurred, even if wastewater leaked from the corroded sewer pipe onto the water tank. All these problems were attributed to poor maintenance on the part of the Management Corporation Strata Title, the system owner, and its managing agent, who had failed to inspect and, where necessary, clean and disinfect the tanks regularly.

Immediate actions were taken by the Public Utilities Board to work with the Management Corporation Strata Title to ensure that the water quality in the contaminated water reticulation system met the WHO *Guidelines for Drinking-water Quality*. These actions included conducting water sampling and cleaning and disinfecting the water tanks and reticulation system and were completed within 4 days of notification of the incident. Instructions were also issued by the then Ministry of the Environment that all food stalls should be closed and that all residents should boil their water. A temporary water supply was provided through water tankers and nearby fire hydrants to ensure continuity of water supply for the residents and office workers during the 4-day period. Water supply was fully restored in the morning of 24 August 2000 after the contaminated tanks and reticulation system were fully cleaned and sterilized. When the water supply was restored, a fibreglass tray was installed beneath the sewage pipe to prevent any future recurrence of such an incident.

To ensure that a similar sewer/water tank setup in developments elsewhere would not lead to a similar incident, an island-wide inspection of approximately 12 000 existing developments was carried out by the Public Utilities Board to verify whether a similar diversion of sanitary pipes was required. Of approximately 12 000 buildings inspected at the point of the incident, some 180 buildings were found to have sewer pipes above the low-level tanks. Of these 180 buildings, 25% had the tanks or pipes diverted; for the remaining 75% where diversion was not possible, fibreglass trays were installed subsequently by the respective Management Corporation Strata Titles as a preventive measure to avoid a future relapse. Each Management Corporation Strata Title was required to engage a professional engineer to carry out annual inspections of the sewer pipes, fibreglass tray, etc. and certify that the sewer pipes were in good condition with no leakage and that the fibreglass tray was intact to collect and channel any leakage of sewage away from the water tank. The then Sewerage Department of the Ministry of the Environment (now the Water Reclamation Network Department of the Public Utilities Board after the merger in 2002) regularly checked to ensure that each Management Corporation Strata Title had submitted the professional engineer's inspection certificate.

Further to the incident in 2000, steps were taken by the Public Utilities Board to prevent a recurrence of such an incident by making changes to the *Public Utilities (Water Supply) Act* in 2001 and the *Public Utilities (Water Supply) Regulations* in 2002 to reinforce and ensure proper maintenance of tanks and to strengthen the Public Utilities Board's enforcement powers against customers failing to maintain their tanks.

The Public Utilities Board also incorporated the key requirements of the provision (the presence of a sewer pipe, floor trap, reclaimed water pipe, waste pipe or any other pipes conveying fluids that may cause contamination of the water in the potable water tanks/low-level tanks located below) into their existing programme of annual inspection and certification of water tanks by licensed water service plumbers. Since 2000, sewer pipes, floor traps, reclaimed water pipes, waste pipes or any other pipes conveying fluids that may cause contamination of the water in the potable water tanks/low-level tanks are not allowed to be installed above the tanks. The mandatory design and installation requirements such as these have also been incorporated in the *Public Utilities (Water Supply) Regulations* and the Singapore Standard CP 48 - Code of Practice for Water Services (SS CP 48) to ensure that water tanks are designed and built to be as contamination-proof as possible. The Public Utilities Board also sends out an annual circular to remind the building owners, Management Corporation Strata Titles and Town Councils about the requirements for regular maintenance of the water service installations.

The various short-term and long-term measures implemented to prevent such a contamination issue from happening again were found to be effective. There has been no relapse of such contamination of the water tanks due to corroding sewer pipes above the tanks since the year 2000.

Linkages to WSP steps

Step 1: Assemble the WSP team
The WSP team included operation and maintenance heads of the then Ministry of the Environment of Singapore, the water supply network department and scientists from the water quality office within the Singapore Water Authority.

Step 2: Description of the system
The water reticulation system within the contaminated building was described.

Step 3: Identification of hazardous events and risk assessment

Various hazardous events had contributed to the contamination, including corroded sewer pipes located above the corroded low-level tank. Wastewater from the leaking sewer pipe seeped through the cracks of the corroded low-level tank and contaminated the drinking-water for the residents within the building.

Step 4: Determination and validation of control measures; Step 5: Develop, implement and maintain improvement plan

Water tanks were immediately cleaned and disinfected. Food stalls were closed, and boil water advisories were issued. Temporary water supply through water tankers and nearby fire hydrants was provided. As a long-term measure, an island-wide inspection was carried out by the Public Utilities Board for approximately 12 000 existing developments to verify whether diversion of sanitary pipes was required. Where diversion was not feasible, a fibreglass tray was installed beneath the sewage pipe. Building owners were required to engage a professional engineer to carry out an annual inspection of the condition of sewer pipes and tanks.

Step 6: Define monitoring of control measures; Step 7: Verification of implementation

Water samples were also collected to verify that the water quality in the contaminated reticulated system could meet the WHO *Guidelines for Drinking-water Quality*. The number of patients suffering from abdominal pain, diarrhoea and vomiting was also monitored.

Step 8: Preparing management procedures; Step 9: Developing supporting programmes

Standard operating procedures were developed and implemented, and training programmes were developed.

Step 10: Periodic review of the WSP; Step 11: Revise the WSP following an incident; Step 12: Enabling environment

Public Utilities (Water Supply) Act, Public Utilities (Water Supply) Regulations, CP 48: Code of Practice for Water Services were reviewed and amended to reinforce and ensure proper maintenance of water tanks and proper design of the customer reticulation system to prevent a recurrence of this incident.

Case-study 9:
Introduction of the "TOKYO High Quality Program" (Tokyo's version of the water safety plan)

A9.1 Features and practical implementation

The World Health Organization (WHO) has advocated water safety plans (WSPs) as a form of comprehensive risk assessment and risk management from water source to tap based on "hazard analysis and critical control points" (HACCP) in order to ensure higher levels of tap water safety.

The Bureau of Waterworks, Tokyo Metropolitan Government, has taken all possible measures to provide safe and good-tasting water for its customers. It has introduced new water quality management measures in addition to conventional water quality management, such as legal examination, monitoring at the water source and during the water purification process, and continuous monitoring through to the water tap. However, its customers' needs for higher water quality continue to increase. To satisfy their needs, water quality management must be strictly controlled. Moreover, as the waterworks supports the functions of the capital Tokyo and its citizens, it has become an urgent matter to implement procedures to prepare for the sudden deterioration of raw water quality and any hazards caused by natural disasters and terrorism.

Therefore, the Bureau of Waterworks has developed and introduced the "TOKYO High Quality Program" as Tokyo's version of WHO's WSPs for improving and maintaining the safety and quality of tap water, not only in the case of emergency, but also in normal operations.

A9.2 Overview of the TOKYO High Quality Program

Along with the WSP based on the HACCP perspective, quality control at all of the water purification plants compliant with ISO 9001 has also been incorporated into the TOKYO High Quality Program. In addition, in order to ensure the safety of tap water to a high degree, the water quality management section of the Bureau of Waterworks has obtained ISO/IEC 17025 certification, which guarantees the objective reliability of its water quality examination techniques.

In this way, the TOKYO High Quality Program is characterized by strict water quality control based on the combination of a WSP based on the HACCP perspective, international standard ISO 9001 and ISO/IEC 17025.

A9.3 Efforts for proper implementation of the programme and practical examples

Under the TOKYO High Quality Program, every related section takes immediate action based on the WSP in the case of an emergency (Fig. A9.1). In normal operation, every water purification plant manages water quality based on the quality manual of ISO 9001, and recording and documentation are also conducted pursuant to ISO 9001. Furthermore, the water quality management section has obtained ISO/IEC 17025 certification, which ensures the objective reliability of the Bureau's techniques for the examination of the quality of tap water.

A9.3.1 Actions for proper implementation of the TOKYO High Quality Program

A9.3.1.1 Education and training

New staff and staff transferred from outside the Bureau of Waterworks go through training on basic knowledge of and skills required for water quality control. More technical education and training are regularly conducted as on-the-job training according to the quality manual.

A9.3.1.2 Data gathering and survey of water quality

The Bureau of Waterworks has been gathering, organizing and utilizing various water quality monitoring data for hazard analysis. The latest data trends have been utilized for review and revision of the TOKYO High Quality Program. Furthermore, the Bureau of Waterworks has been gathering information on the usage of harmful substances and potentially hazardous or unregulated substances in water catchment areas.

A9.3.1.3 Information service for customers

To help obtain customers' trust in their tap water, information related to the TOKYO High Quality Program has been actively disclosed to the public in an easily and clearly understood manner. This information is on water quality control from water source to tap, how such quality control has been implemented, and responses to possible future hazards.

A9.3.2 Practical examples

Sixty-five practical cases were reported and responded to, based on the TOKYO High Quality Program, between April 2012 and March 2013. Those cases included water quality deterioration due to the operation of a drainage pump station, a sudden increase in turbidity as a result of typhoons and torrential rains, a musty odour occurrence due to the growth of algae and improper discharge of industrial wastewater. Noteworthy cases over this period were excess formaldehyde concentrations in finished water due to the discharge of industrial wastewater that contained a formaldehyde precursor and cross-connection of the pipe for rainwater utilization to the customers' service pipe. Each case was appropriately responded to – by increasing the amount of polyaluminium chloride injected, injecting powdered activated carbon, changing the allocation of water distribution among water purification plants and other countermeasures.

A9.4 Summary

WSPs, the new water quality control approach advocated by WHO, have enabled the Bureau of Waterworks to make a transition from the perspective of emphasizing water quality inspection to one emphasizing process management. Furthermore, an integrated water quality inspection method with higher accuracy has been introduced to the TOKYO High Quality Program – that is, risk management based on WSPs and quality management at every water purification plant based on ISO 9001 and ISO/IEC 17025. In this way, Tokyo's version of the WSP has been developed with a unique perspective. Along with continuing implementation and improvement of the TOKYO High Quality Program, the efforts of the Bureau of Waterworks and the implementation status of the TOKYO High Quality Program will be actively shared with the Bureau's customers. The satisfaction of the Bureau of Waterworks' customers and their trust in their tap water will be further secured through those activities.

Abnormal turbidity and related items in distribution network
*related items: iron, manganese, odor, color, foreign matter

Possible Cause	Distribution	- Occurrence of turbidity due to aging of pipes, failure of power supply, or other incidents
Confirming the incident		**Monitoring: Automatic turbidity meter/chromatometer, and routine water quality inspections** **(1) Detecting abnormal turbidity** ☐ Check abnormal turbidity by value of automatic turbidity meter/chromatometer in the distribution area. ☐ Check abnormal turbidity by value of related items observed by routine water quality inspection in the distribution area. **(2) Confirming accuracy of the monitoring instruments and reconfirming the results of water quality inspection** ☐ Reconfirm abnormal turbidity by value of automatic turbidity meter/chromatometer in the distribution area. ☐ Reconfirm turbidity and related items in the preserved same water samples. ☐ If the abnormal turbidity was confirmed again, conduct emergency water quality inspection at outlet of water purification plants, distribution reservoirs, other automatic turbidity meters and hydrants in the distribution area. **If the values of water quality inspection are normal and the abnormal value was due to the monitoring instrument error, adjust the instrument and see how it works.** **(3) Confirming affected distribution area, determining the risk level and identifying the origin of the abnormal turbidity if the abnormal turbidity was finally confirmed** ☐ Identity distribution area affected by the abnormal turbidity by the result of water quality inspection, and determine the risk level. ☐ Identify the origin of the abnormal turbidity such as nearby construction works. ☐ Notify the relevant departments in the bureau. **In case the abnormal turbidity comes to water tap, also conduct countermeasures of "Abnormal turbidity in tap water"**
Countermeasures		**Risk level 3: turbidity and/or related items are likely to exceed drinking water quality standard** (1) Countermeasures at water stations ☐ Consult with the relevant departments in the bureau and mix purified water from other networks at water stations. ☐ Step up monitoring of turbidity and other related items at water stations, and confirm absence of abnormal water quality. (2) Countermeasures in water distribution networks ☐ Refer to administrative map of water distribution network and conduct drain from drainage facilities and/or hydrants. ☐ Step up monitoring of turbidity and other related items at hydrants in the affected distribution area, and confirm absence of abnormal water quality.

Fig. A9.1. Examples from the emergency response manual of the Bureau of Waterworks, Tokyo Metropolitan Government

	Risk level 5: turbidity and/or related items have exceeded drinking water quality standard
Countermeasures	(1) Countermeasures at water stations ☐ Stop water distribution. Consult with the relevant departments in the bureau and conduct emergency drain and receive purified water from other networks. ☐ In case the purified water from other networks is not available and water stoppage/shortage is unavoidable among the affected distribution area, conduct emergency water supply by water trucks and emergency information activities, cooperating with the relevant departments. ☐ Conduct monitoring of turbidity and other related items at water stations, confirm absence of abnormal water quality, and resume water distribution. (2) Countermeasures in water distribution networks ☐ Stop water distribution and conduct allocation adjustment among water distribution networks in order to avoid the spread of the abnormal turbidity. ☐ Consult with the relevant departments in the bureau and receive purified water from other networks. ☐ In case the purified water from other networks is not available and water stoppage/shortage is unavoidable among the affected distribution area, conduct emergency water supply by water trucks and emergency information activities, cooperating with the relevant departments. ☐ Refer to administrative map of water distribution network and conduct drain from drainage facilities and/or hydrants. ☐ Conduct monitoring of turbidity and other related items at hydrants in the affected distribution area, confirm absence of abnormal water quality, and resume water distribution.

Fig. A9.1 (continued). Examples from the emergency response manual of the Bureau of Waterworks, Tokyo Metropolitan Government

Linkages to WSP steps

Step 1: Assemble the WSP team

The WSP team included operation and maintenance heads of the Bureau of Waterworks, Tokyo Metropolitan Government.

Step 3: Identification of hazardous events and risk assessment

Water quality monitoring data were gathered and used for hazard analysis.

Step 8: Preparing management procedures

Standard operating procedures on water quality control were developed and implemented as part of the quality manual.

Step 9: Developing supporting programmes

Training on basic knowledge and skills of the water quality control programme was developed, and the programme was mandatory for all new staff and staff transferred from outside the Bureau of Waterworks. A technical education and training programme was also developed as part of the quality manual.

Case-study 10:
Incidents from "Lessons learnt from plumbing incidents – responses and preventions", Japan Water Plumbing Engineering Promotion Foundation (2011)

A10.1 Incidents of cross-connections with industrial water pipes at branches

A10.1.1 Incident 1-1: Cross-connection of a distribution sub-main to an industrial water pipe

[Background to detection] There was a dysentery outbreak at a kindergarten in City Y in June 1969. Investigation showed no detection of residual chlorine at the outlets of faucets. Water to this kindergarten was supplied by a distribution sub-main with a diameter of 50 mm, through which industrial water flowed.

[Cause of accident] To supply water to households in this area, feeder pipes were branched from a 50 mm diameter distribution sub-main branching from a 150 mm diameter distribution main. The 150 mm distribution main had to be removed because it was obstructive to sewerage works. Accordingly, the 50 mm distribution sub-main had to be reconnected to a 200 mm distribution main. During the connection work, the sub-main was erroneously connected to a 200 mm industrial water pipe, which was laid parallel to tap water piping.

[Background to accident] The distribution main could not be distinguished from the industrial water pipe of the same diameter because there was no unified piping ledger prepared for the maintenance of feeder and distribution pipes, and residual chlorine concentration was not checked during the branch construction.

A10.1.2 Incident 1-2: Cross-connection associated with new feeder pipe installation

[Background to detection] A water leakage incident occurred in August 2002 on a road within the supply area of the waterworks bureau of City T. During the repair work, a feeder pipe connected to an industrial water pipe was found. The branch construction was conducted in July 1996, which means that industrial water had been supplied to households for 6 years. The industrial water had been supplied without chlorination after the settlement treatment of river water.

[Action taken] The waterworks bureau of City T immediately changed the connection of the feeder pipe to a tap water supply system and apologized to the affected household residents, who were then subjected to health inspection. Fortunately, no health effects were found.

[Background to accident] The water supply pipe and the industrial water pipe had the same colour and diameter and were laid in parallel, so that they were prone to confusion. Moreover, the person responsible for the branch construction failed to identify the pipe carefully, and residual chlorine was not checked for during the branch construction.

A10.1.3 Incident 1-3: Cross-connection at a branch during lead pipe replacement work

[Background to detection] This accident was revealed in 2006 by a complaint from an office in City O that white water came out of a tap. A contractor who conducted lead pipe replacement erroneously connected a feeder pipe to an industrial water pipe, which was laid parallel to a tap water pipe. As a result, industrial water, instead of tap water, had been supplied to the office for about a half year.

[Background to accident] Although a tap water pipe of 50 mm in diameter and an industrial water pipe of 75 mm in diameter were specified on the management diagram of the relevant underground facility, the actually laid pipes were a 25 mm tap water pipe and a 50 mm industrial water pipe. The contractor mistook the 50 mm industrial water pipe as a tap water pipe and made an erroneous connection. City O did not check for residual chlorine at the completion of the work.

> **Lessons learnt:**
>
> - Because industrial water is purified to some extent, it is difficult to detect water quality anomalies in this type of accident. Once connected, because this type of cross-connection is difficult to find, industrial water tends to be supplied to households for a long period of time. Investigation should therefore be performed carefully before and after the branching construction.

A10.2 Incidents of other cross-connections at branches

A10.2.1 Incident 2-1: Backflow of construction site wastewater into a distribution pipe

[Outline of accident] This accident occurred in 1991. There were complaints from a residential area about "something like oil floating on tap water". In response, the waterworks bureau immediately used a hydrant to drain the turbid water and cleaned the distribution pipe, used a loudspeaker van to advise the residents not to drink the tap water and used a water truck for an emergency water supply. At the same time, the bureau investigated the cause of the accident. As a result, a reversely rotating water meter was found at a nearby shield driving construction site, and the material detected in the contaminated tap water was found to be the same as the oil (used for the shield machine) contained in the wastewater generated at the construction site. This accident affected about 2000 households and resulted in the replacement of about 2800 water meters due to oil adherence and other problems.

[Background and cause] At the construction site, wastewater generated in the tunnel was being stored in a wastewater pit in the vertical shaft, pump 1 in the pit was being used to pump up the wastewater to a settlement tank on the ground, and, after turbid water treatment, the wastewater was being discharged to the sewerage system. In contrast, because there was clear spring water in the shaft, a drum was set in the wastewater pit to collect the spring water, and after pumping up the spring water with pump 2 to a settlement tank on the ground, the spring water was being stored in a receiving tank and pressurized with a water supply pump for reuse as construction water. Because spring water was not enough for the demand, tap water was fed to the receiving tank. In addition, to allow tap water to be used directly as construction water without going through the receiving tank, a non-qualified person installed a bypass pipe that directly connected the feeder pipe with a pipe located downstream of the receiving tank.

On the day of the accident, pump 1 operated when a worker was cleaning the wastewater settlement tank, and turbid water entered the tank. The worker therefore switched off pump 1, but mistakenly thought that pump 2 was also switched off. Moreover, the worker opened the bypass pipe because of concern that the spring water would not be supplied and construction water would become insufficient. As a result, the level of turbid water in the wastewater pit rose; the turbid water flowed into the drum for collecting spring water and was supplied to the receiving tank through pump 2. Because the discharge pressure of the water supply pump was higher than the pressure of the distribution pipe, the turbid water flowed backward to the distribution pipe and was supplied to houses around the construction site (see Fig. A10.1).

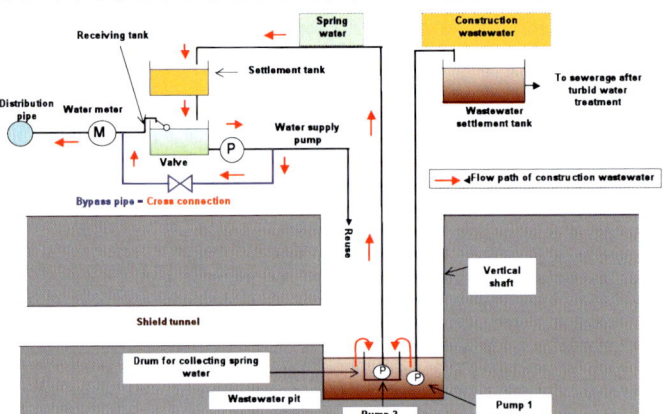

Fig. A10.1. Outline of a backflow accident at a construction site

[Action taken] The waterworks bureau ordered the removal of the bypass pipe. Since the accident, in order to prevent accidents, the waterworks bureau has been conducting periodic on-the-spot inspections at all the construction sites where construction water is being supplied.

A10.2.2 Incident 2-2: Backflow of agent from bactericide application system to feeder pipes

[Outline of accident] In an accident that occurred in 2007, residents notified that tap water was yellowish and odorous. It was revealed that a pipe of a bactericide application system in an orchard was directly connected to a feeder pipe; the bactericide flowed backward to the feeder pipe and flowed out of water taps in three nearby houses.

[Action taken] When notified of the water quality anomaly, the waterworks bureau immediately advised the users of water supplied from the affected feeder pipe not to use the tap water and provided an emergency water supply using polyethylene tanks. It also drained the feeder pipe, conducted a water examination to find that the anomaly was due to bactericide and identified that the orchard using the same contaminated feeder pipe was the source of contamination (see Fig. A10.2).

The bureau cleaned the feeder pipe, verified its safety and then notified the users that the tap water was drinkable.

[Cause and background] This accident, in which a feeder pipe was directly connected to a bactericide application system in an orchard, was caused by insufficient understanding by the bactericide application system installer and the orchard owner of Japanese Structural and Material Standards for water service installations.

Fortunately, because the affected residents immediately noticed the anomaly and stopped using tap water and notified the waterworks bureau of the anomaly, the impact of the accident was minimized.

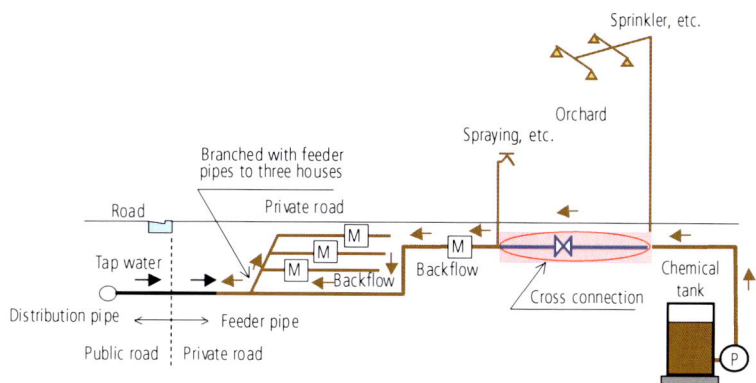

Fig. A10.2. Outline of a backflow accident at an orchard

A10.2.3 Incident 2-3: Backflow of industrial water to feeder and distribution pipes

There was a report from a tap water user of City O complaining about smell and turbidity in tap water. A water inspection revealed a free residual chlorine concentration less than 0.1 mg/L.

To identify the cause of the water quality anomaly, an inspection of the tap water was conducted. It was found that the measurements of chloric acid, bromate and trihalomethanes were different from those of the city's tap water. There were two factories using industrial water in the relevant area; therefore, the pathway of industrial water contamination was investigated. As a result, it was found that a feeder pipe was connected to an industrial water pipe in a factory to which water was supplied through a distribution pipe laid in the relevant area. Because of a water pressure difference at night, industrial water flowed backward from the factory to the distribution pipe.

A10.2.4 Incident 2-4: Outflow of foreign matter from a water tap through a well water pipe

In response to a report that a worm flowed out of a lavatory faucet of one of the two houses built on the same lot, an on-the-spot survey was conducted. Water did not stop flowing even when a check stop located upstream of the meter was closed, and it was found that a tap water pipe was connected to a well water pipe.

The house was built on a site where there was once a barn for storing farm tools and other equipment and materials for home gardening. For the main source of domestic water, a water service installation for the existing house was additionally installed (modification work). The existing well water, which was intended for a pond, spraying and bath, was connected to the water service installation (see Fig. A10.3).

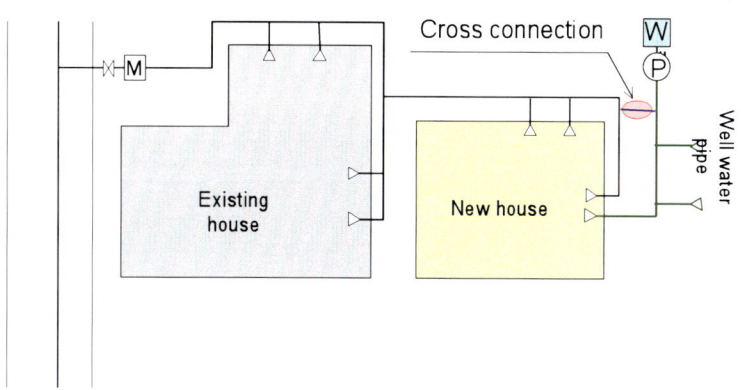

Fig. A10.3. Backflow of well water containing foreign matter

> **Lessons learnt:**
>
> - When supplying tap water through a pipe in which other liquid or gas flows, it is necessary to install a receiving tank on the tap water side, secure an air gap by using a ball tap or the like and make a connection in isolation from the feeder pipes. When pressure is required on the tap water side, a pump should be installed downstream of the receiving tank.
> - The second most frequent type of cross-connection is connection to well water pipes. A request for this type of piping often comes from a client. The chief engineer of water service installation works must decline the work offer by explaining to the client that this type of connection causes contamination of tap water and is against the Waterworks Law in Japan.

A10.3 Incidents related to construction work

A10.3.1 Incident 3-1: Leakage associated with meter replacement

After replacing an anomalous meter, water leaked from the clamping part of a nut, and an underground machine room was flooded (see Fig. A10.4). The causes of the accident were as follows:

- Feeder pipes around the meter were corroded.
- The nut was not tightened sufficiently at the time of meter replacement.
- The nut was loosened by some impact after the construction work.

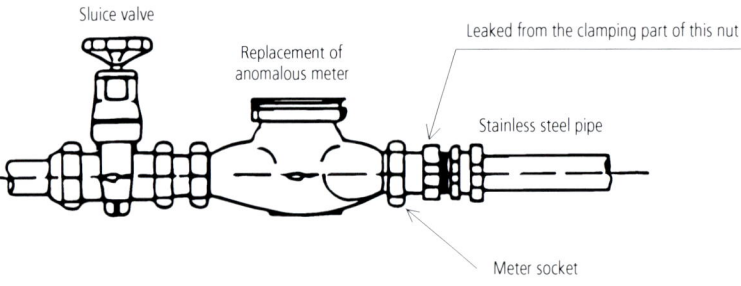

Fig. A10.4. Leakage at the time of meter replacement

A10.3.2 Incident 3-2: Leakage from a corroded corporation stop with saddle

Four years after installing a corporation stop with saddle (100 mm/25 mm in diameter) in 1995, corrosion and leakage occurred around the main body. Investigation showed that a stainless steel pipe was connected to the branch side without an intermediate insulator, and therefore bimetallic corrosion due to contact between two types of metal with different potentials occurred.

> **Lessons learnt:**
>
> - Possible causes of those accidents include: (1) insufficient knowledge on the structural and material standards, (2) immature construction skills and (3) insufficient knowledge on the performance, functions, structure and other aspects of various feeder pipes and water service devices. If an accident is revealed after the completion of construction, it may require significant efforts and costs for recovery.

A10.4 Penetration of organic solvents into synthetic resin pipes

A10.4.1 Incident 4-1: Odour caused by oil leakage due to kerosene pipe corrosion

A user complained about an oily odour in water. A kerosene-like odour was detected in water, faucets and drain tap packing at the user's place. Investigation showed that oil leaked due to corrosion of a kerosene pipe that penetrated into a polyethylene pipe laid parallel to the kerosene pipe. Soil had to be replaced, and the polyethylene pipe had to be replaced with a copper pipe (see Fig. A10.5).

Although the direct cause of the incident was oil leakage due to the kerosene pipe corrosion, the fact that the feeder pipe was laid close to the kerosene pipe was also a problem. Chief engineers of water service installation works must explain this kind of risk to house owners and building companies and take precautions to prevent accidents.

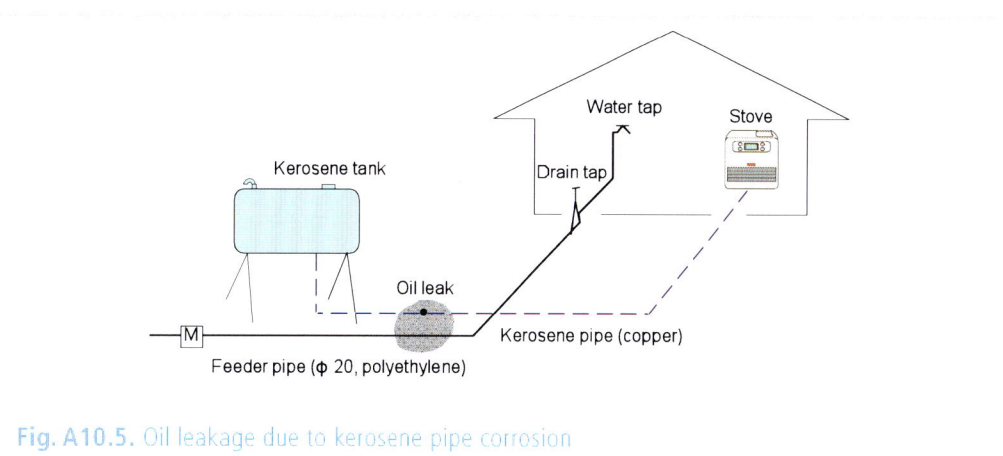

Fig. A10.5. Oil leakage due to kerosene pipe corrosion

A10.4.2 Incident 4-2: Odour due to an organic solvent dumped by a painter during house building

A resident of a new house complained about an odour of paint thinner in tap water. As a result of investigation, an odour of paint thinner was detected in an area where a feeder pipe was buried.

A painter inadvertently dumped an organic solvent used to thin paint on an area under which a polyethylene feeder pipe was buried, and the solvent penetrated into the pipe, causing the odour of paint thinner in the water.

A10.4.3 Incident 4-3: Odour due to penetration of herbicide

When water started to be supplied after a feeder pipe was installed, there was a complaint about a strong oily odour. Investigation showed that herbicide was sprayed on an area where underground piping was laid, and this herbicide penetrated into sandy soil and a buried polyethylene pipe, causing an odour in the tap water.

Lessons learnt:

- Synthetic resin pipes (polyethylene, polyvinyl chloride, bridged polyethylene and polybutene pipes), which are resistant to corrosion and adaptable to construction work, are widely used for feeder pipes. A weakness of synthetic resin pipes is that they swell and rupture when exposed to organic solvents, and penetration of organic solvents into the pipes results in pollution of tap water. In the vicinities of business institutions that handle gasoline, kerosene, thinner and the like, and in areas where kerosene tanks are used for heating, it is necessary to use alternative types of pipes or protect synthetic resin pipes by inserting them in casing pipes.
- Painting is often conducted after laying feeder pipes in the case of building a new house, and similar accidents may occur if there is a synthetic resin pipe buried under an area where thinner or other liquid used for washing coatings and brushes is dumped. Those involved in the construction work should be warned of this possibility.
- When termite repellents, pesticides, herbicides and the like containing volatile substances are sprayed and come into contact with a synthetic resin pipe, they may penetrate into the pipe. It is therefore necessary to take measures similar to the above-mentioned ones in areas where such materials may be sprayed, such as an area below the floor.

A10.5 Incidents of sandblasting causing damage to other underground facilities

A10.5.1 Incident 5-1: Gas supply suspended by inflow of water and soil to a gas pipe

Water leaked from a polyethylene feeder pipe (20 mm in diameter), and tap water entered a steel gas supply pipe (50 mm in diameter) laid immediately below this feeder pipe (about 10 cm in distance at the crossover), which was perforated by sandblasting. As a result, gas supply to about 200 houses was suspended (see Fig. A10.6).

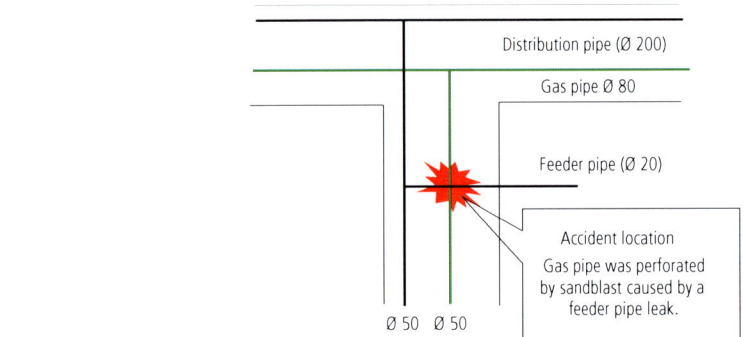

Fig. A10.6. Plan view showing the location of a damaged gas pipe

The accident occurred in an area of high pressure (0.75 MPa). It is considered that polyethylene resin was deteriorated by chlorine contained in tap water after a long time of use, cracks were developed due to a water hammer that acted over a long time period, and water leaked from the feeder pipe. This feeder pipe, branching from a private pipe laid for housing development, had an insufficient distance to the gas pipe.

The water utility negotiated with citizens and the gas company concerning compensation for the accident and for the damage to the gas company and other parties.

> **Lessons learnt:**
>
> - Accidents related to sandblasting or sand erosion are most frequently reported as damage to gas pipes. This probably reflects that water and gas lines are buried under roads, which are indispensable for daily life and are branched to residential houses mostly at the same location and depth; water feeder pipes are therefore laid about 300 mm or more distant from gas pipes to prevent accidents.